NORTHEASTER

A STORY OF COURAGE AND SURVIVAL
IN THE BLIZZARD OF 1952

CATHIE PELLETIER

PEGASUS BOOKS
NEW YORK LONDON

NORTHEASTER

Pegasus Books, Ltd.
148 West 37th Street, 13th Floor
New York, NY 10018

First Pegasus Books cloth edition January 2023

Interior design by Maria Fernandez

Map on page xvii by Lara Taber

Library of Congress Cataloging-in-Publication Data is available.

ISBN: 978-1-63936-341-4

10 9 8 7 6 5 4 3 2

Printed in the United States of America
Distributed by Simon & Schuster
www.pegasusbooks.com

To the memory of

HARLAND DAVIS AND JAMES HAIGH

HAZEL TARDIFF

RAY "SONNY" POMELOW

JAMES MORRILL AND PETER GODLEY

PAUL DELANEY

GEORGE ASPEY

CHARLES VOYER

WILLIAM DWYER

And to their family members, some of whom answered hundreds of questions with willingness and patience, and supplied necessary documents. They did this during two trying years of their own lives with COVID-19. Without them, there would have been no book.

Mary Tardiff Wirta

Dennis and David Tardiff (children of Hazel Coombs Tardiff)

William "Bill" Wilson (son of Harland and Alice Davis)

Rebecca Sue Godley (daughter of Peter Godley)

Barbara "Barbi" Haigh (daughter of James and Eleanor Haigh)

Kathleen Delaney McNamara (daughter of Paul V. Delaney)

Gerry and David Gamage (sons of Weston Gamage)

CONTENTS

PART FIVE

PART SIX

AUTHOR'S NOTE

Most people who know of the northeaster of February 17–18, 1952, will likely associate it with the loss of the SS *Pendleton* and the SS *Fort Mercer*. These two large ships had been sailing northward to New England when the storm overtook them. Tankers the size of football fields, they were running late thanks to ragged seas with sixty-foot waves and gale winds of eighty miles an hour. On the morning of February 18, the second day of the storm, each ship cracked in two and floundered twenty miles apart off Cape Cod. The stories of their rescues and losses have made coast guard history. A film, *The Finest Hours,* tells the heroic tale of the SS *Pendleton.*

In *The Story of Civilization*, Will Durant wrote that "Civilization is a stream with banks. The stream is sometimes filled with blood from people killing, stealing, shouting and doing the things historians usually record, while on the banks, unnoticed, people build homes, make love, raise children, sing songs, write poetry . . . The story of civilization is the story of what happened on the banks."

This book is also a story of what happened *on the banks* during one particular storm in 1952. It was far from the most dramatic storm in New England's history. But it's the drama that happens to ordinary people, to ordinary families, to ordinary towns—stories often not press-worthy—that become footnotes, if not chapters, to that bigger saga of civilization. It's the stuff novels are most often built from: the commonplace of everyday lives when disrupted. What rises from the chaos deserves remembering.

The wind may blow the snow about,
For all I care, says Jack,
And I don't mind how cold it grows,
For then the ice won't crack.

Old folks may shiver all day long,
But I shall never freeze;
What cares a jolly boy like me
For winter days like these?

—"A Country Boy in Winter" by Sarah Orne Jewett

THE FIRST STORM

Come see the north wind's masonry.
Out of an unseen quarry evermore
Furnished with tile, the fierce artificer
Curves his white bastions with projected roof
Round every windward stake, or tree, or door.

—"The Snow-Storm" by Ralph Waldo Emerson

t was early on Monday morning, February 11, 1952, that it started snowing hard over the timberlands in northern Maine. Aroostook County, the largest and most sparsely populated, was hit hard, with Portland and smaller towns three hundred miles to the south receiving only cold rain. Northern Maine was a different world to the Maine most summer people knew. Hours from the ocean, its little communities thrived on rivers and lakes, in a world built on potatoes and lumbering. Even the tourists who trekked that far north were of a more rugged mindset. They were canoers and naturalists whose idea of fine dining was trout fried over a riverbank fire, with biscuits baked in a pan nestled on hardwood coals. The lobster was still an alien creature that far north. And nobody born and raised in Aroostook County ever said, "A-yuh, good mawnin' this mawnin'." That's how people who lived in the southern part of the state talked, at least in Hollywood's version of Maine.

By the time Tuesday dawned to gray skies, thirty inches of snow had fallen in just over twenty-four hours. As town plows went to work and small communities along state highways were digging themselves out, a call for help reached the northern town of Patten. It was made using a portable telephone connected to a line the Diamond Match Company had strung fifteen miles into the forest east of Baxter State Park. Diamond needed wood for its matches and toothpicks, and a telephone line would facilitate communication with workers at the company's camps. Plodding over snow that had piled shoulder-deep, two lumberjacks on snowshoes had managed to reach the telephone line, fifteen miles from their camp at Second Lake Matagamon. They reported that more than a hundred people, thirty-one horses, and seventeen head of cattle were stranded at three lumber camps operated by Diamond and the Eastern Paper Corporation. The workers had been expecting a truck loaded with their weekly provisions on Tuesday. But then the snows came and blocked the tote road. Now they were marooned in a wilderness of trees. Supplies were running low and the animals were down to their last meal.

While thirty inches of heavy snow would have strained ordinary men and tested modern equipment, it could still be managed in rugged northern Maine. But if a persistent wind got involved the situation could change quickly. *Drifting snow* can create piles and banks up to eight feet high. Any taller than that and it becomes *blowing snow*, which is carried higher on the winds and greatly reduces visibility. Even with snow no longer falling, strong gusts of wind can continue to deliver havoc, especially if it's intent on blowing in one direction for several hours or days. And that's what the wind did to the freshly fallen snow in northern Maine. It blew over it nonstop for four days.

Now snow blocked the solitary twenty-seven-mile tote road that led into the camps, the last section of which had been given the nickname "Burma Road" by the woodsmen who walked it. No matter how much effort the plow-equipped trucks and a bulldozer affectionately named "Alice" had put into opening the road back up, the blowing snow refilled it. According to some reports, the drifts were fifteen to twenty feet high.

As was customary with the operation of a lumber camp, the workers had gone to the camps and settled in before the winter snows came. With them were six women, some accompanying their husbands and bringing small children. They would cook for the operations all winter as the men cut timber for their employers. The crews wouldn't come back out until the spring thaw. That crude tote road kept them supplied weekly when a company truck rumbled in with crates of eggs, fifty-pound bags of flour, burlap sacks of potatoes, coffee, powdered milk, canned vegetables, pounds of beef, and boxes of frozen fish. They also brought hay and grain for the animals. But now all entry to the camps had become blocked since Tuesday. No fresh supplies had been delivered for a week. A few more days and the situation would become serious. Even hunting for deer or rabbits in blowing snow was near impossible. And the work animals were down to their last bales of hay and already feeling the strains of hunger.

An emergency call went out to the small municipal airport in Millinocket, thirty miles below Patten and sixty miles south of the Diamond woods camp at Second Lake Matagamon. Three bush pilots immediately fueled their small planes that were already equipped with wide wooden skis for the season. These were experienced men, accustomed to landing on snow-packed terrain during the worst weather and the poorest visibility. Their rescue plan was christened Operation Snowbound. Those stranded workers in the three remote camps would soon hear a plane's engine buzzing above the clouds of swirling snow.

The first pilot banked his Cessna near the larger camp operated by the Diamond Match Company. Swooping as low as he dared above the blowing gusts, he dropped a side of beef and watched it spiral down into a snowbank below. A second plane dropped more food, and a third hay and grain for the horses and cattle. Over the next two days, these veteran pilots managed to land on nearby lakes and ponds, and on makeshift runways packed down by lumberjacks wearing snowshoes. With them, they brought supplies that couldn't be dropped from the air.

The hours wore on and still the wind blew snow into towering drifts. That one road out of the deep woods remained blocked. By the time Friday night rolled around the workers and animals had not been rescued. While they had enough food for the time being, the situation looked grim. The U.S. Weather Bureau down in Portland was now predicting "a fairly heavy northeaster" to begin Sunday morning and be over by Tuesday. This forecast again targeted northern Maine, troublesome news for the lumber camps. People who lived and worked in nature had learned long ago that nothing is certain when storms appear on the horizon. Placing a bet on any weather forecast could be gambling with lives. What if the storm lasted longer, and again brought blowing snow? Time was now of the essence. The plow trucks had been steadily working, whittling away at snow that the wind tossed back onto the road once a plow had passed. The drifts now towered twenty feet in the less sheltered stretches.

On Saturday morning the sun was shining. The wind curled back into the woods and lay dormant. Just as talk began of evacuating the women and children by airlift, word came that the truck plows and Alice the bulldozer had broken through. The tote road to the two Diamond camps was now open. At the sight of an approaching truck, the first one that camp workers had seen in almost two weeks, joyful shouts welcomed the drivers. Rescue vehicles from Patten were following the plows. The women and children climbed into warm automobiles. Lumberjacks rode on plows or jumped onto the backs of pickup trucks. Many of them walked, their red-and-black woolen jackets an undulating ribbon against the banks of white snow. Horses and cattle labored along behind, while the weaker animals rode on trucks. Orderly and calm, the caravan of humans, animals, and vehicles began its journey out of the woods. Patten, the nearest town, was thirty miles away.

This snowstorm, with its blizzard-like conditions, had moved east from the Great Lakes, pushing a warm front ahead of it. It had gone out with a satisfactory ending. But a northeaster was just picking up its power off Cape Hatteras on the Outer Banks of North Carolina. It was moving north along the coastline. Before this incoming storm would be over, three dozen people in New England would be dead, six of them in the state of Maine.[1]

THE PRINCIPAL PLAYERS

HAZEL TARDIFF, thirty-four, housewife, Bath; expecting her fourth child.

HARLAND DAVIS, thirty, lobster fisherman, Pleasant Point.

JAMES HAIGH, thirty-nine, lobster wholesaler, Portsmouth, New Hampshire.

RAY "SONNY" POMELOW, fifteen, high school student, Brownville.

JAMES MORRILL, thirty, bartender, Brewer.

PETER GODLEY, thirty, shoe factory worker, Brewer.

BILL DWYER, sixty-seven, retired worker and fireman, Bath.

CHARLES VOYER, sixty-three, retired theater worker, South Portland.

GEORGE ASPEY, fifty-six, woolen mill carder, Warren.

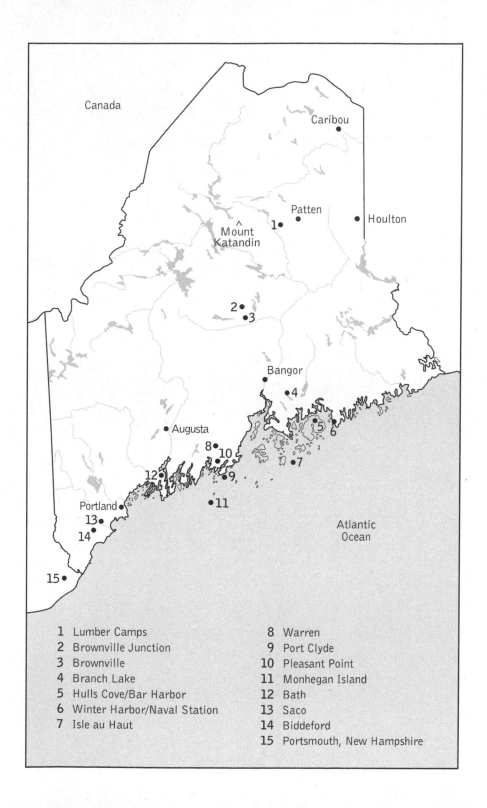

Canada

Caribou •

Patten
1 •   • Houlton
^
Mount
Katandin

2 •
  • 3

Bangor
• 4

• Augusta

8 •
  10 •
12 •   9 •   • 7
           • 11

Portland •
13 •
14 •

15 •

Atlantic
Ocean

5
6

1 Lumber Camps   8 Warren
2 Brownville Junction   9 Port Clyde
3 Brownville   10 Pleasant Point
4 Branch Lake   11 Monhegan Island
5 Hulls Cove/Bar Harbor   12 Bath
6 Winter Harbor/Naval Station 13 Saco
7 Isle au Haut   14 Biddeford
                                15 Portsmouth, New Hampshire

PART ONE

FEBRUARY 16, SATURDAY

*It was difficult to know this country without having wintered
there; for on arriving in summer everything is very pleasant on
account of the woods, the beautiful landscapes, and the fine
fishing for the many kinds of fish we found there. There are six
months of winter in that country.*

—Samuel de Champlain's journal, 1604,
regarding Saint Croix Island, Maine

BATH, MAINE: THE CITY OF SHIPS

A year after that winter at Saint Croix Island, Samuel de Champlain
sailed from the Gulf of Maine and up the Kennebec River, named by
the indigenous Abenaki people to mean "long level water." In 1781,
nearly two centuries after Champlain's curiosity in the Kennebec, a settlement
had developed on the northern end of a long and narrow peninsula. It became
incorporated that year and eventually spread out parallel to the river. Fifteen
miles inland from the coast, Bath, Maine, settled down to a long history of
fishing and shipbuilding, with shipyards dotting the riverbanks. Known for

its accessible harbor, many weary sailors appreciated not having to change a ship's rigging when sailing into Bath's welcoming port.

A few years before the town would construct buildings designed in the then popular Greek and Italian architecture styles, Bath would know a turbulent and less romantic time. The most violent moment in its history occurred in the summer of 1854 during an uprising of anti-Catholic, anti-immigration, and anti-foreigner sentiment that was sweeping the nation. The American Party was at the vanguard of this unrest. They had been nicknamed the Know Nothing Party since they replied "I know nothing" when questioned.

In Bath it began when an itinerant preacher named John Sayers Orr turned up one day dressed in a white robe and carrying a trumpet that he played on a street corner to attract a gathering. Once people had assembled, the "Angel Gabriel," as Orr called himself, had the unique ability to transform enough of them into a seething, anti-Catholic mob. He had just succeeded in performing this hateful feat in New York City and Boston. Now he was in Mid-Coast Maine.

On July 6, 1854, on a street corner in Bath, the blasts from John Orr's trumpet quickly summoned a rapt crowd. Some newspaper reports claimed that its numbers swelled to "over a thousand," an eighth of the population. Stoked to fury from the Angel Gabriel's anti-Catholic vitriol, the throng advanced on the Old South Church, on High Street. This house of worship had been constructed in 1805 and was later purchased by Irish Catholics. The rioters hoisted above the belfry an American flag boasting thirty-one stars. Then they set the church on fire. At dawn, with nothing left but crashed rafters and cinders, the lingering mob finally dispersed. Mission accomplished, the Angel Gabriel packed up his trumpet and crossed the ocean to incite more riots in Scotland and later England.[2]

When talk of the episode finally died down, Bath concentrated again on shipbuilding. In 1884, Bath Iron Works was founded. That shipyard would eventually employ a multitude of local workers for generations to come. The company would build hundreds of wooden and steel ships. It was estimated that during the Second World War, that one yard launched a new ship every seventeen days. The city prospered.

THE TARDIFF FAMILY OF BATH

Hazel Tardiff sifted flour for the biscuits she was making for supper. She added baking powder, salt, shortening, and then poured milk into the mixture. A pot of beans and molasses, with a chunk of salt pork, was already in the oven. Before she cut the rolled dough, she turned on the floor-model radio Phil had put in their kitchen. She often teased him that he had chosen brown to match their wood paneling. She rolled the dial to WPOR in Portland. There were lumberjacks, including women and children, snowed in up north. Hazel knew she wasn't alone in praying for their safety each night. The local newspapers had been filled with updates, how airplane pilots had dropped food and even hay for the animals. The papers had also predicted that another storm was on the way, this time a northeaster that was expected to again target northern Maine.

Biscuits ready for the oven, Hazel straightened, pressing fingers against her lower back. She was thirty-five years old and this was her fourth pregnancy. It hadn't been a difficult one so far, but winters were long enough in Maine as it was. Since Christmas her activities had become more and more limited. She was ready for it to be over. Boy or girl didn't matter, but a boy would make it even. Two girls, two boys. Dr. Virginia Hamilton expected the baby would come sometime within a week. "It could be any day," she said. "Maybe you should move to town just to be safe." The Tardiff home, a brick Cape built in 1850, sat four miles north of Bath, near the left bank of the Kennebec River. Scattered with wildflowers each summer, the hayfields surrounding the house were now covered in snow. Their country road, Varney Mill, had to be plowed regularly or the few families who lived out there would be cut off from town. Hazel decided to take it one day at a time.

Hazel Coombs Tardiff didn't always live in Bath.[3] She was born and raised on Isle au Haut, a small island in Penobscot Bay that Champlain declared as "High Island" during his 1604 voyage, a year before he explored the Kennebec. Accessible only by the mailboat from Stonington, a forty-five-minute ride away, the island had never anticipated a large population. Only six miles long and two miles wide, it was content to remain a world apart from the mainland with its fifty Scottish and English families making a living by fishing, farming, and raising sheep. At the end of the 1800s, the census report listed 275 villagers as permanent residents, the Coombs family among them.

Before that century turned, visitors from cities like Boston and New York—the locals called them *rusticators*—began spending their summers on the island. Later, as a teenager, Hazel had worked as a waitress in the private club at Point Lookout where the wealthy summer people built their cottages. She and the other girls wore a maid's uniform, a starched full bib apron over a dark blue dress, crisp white hat, and starched white cuffs. That income was needed money. Some mornings the family had only lobster or fish to eat for breakfast. There was one winter when her father, George Coombs, had a nickel left in his trouser pocket when spring finally came.

Born in 1917, Hazel grew up a short distance from Coombs Mountain, named for the first family member to settle on the island. She had gone to a one-room grammar school and attended the community's one church. For entertainment, there were dances held at the town hall. Hazel loved the island. It had forests thick with conifers where her father, a fisherman, hunted for game to help feed his family. Deer and rabbit were a welcome change from the lobster and cod he took from the sea. There were low-lying hills and even meadows that bordered the forest. Granite boulders that the eons had sculpted into smooth, rounded shapes hugged the coastline. There were cormorants, sleek seals, and loons calling from the hidden coves.

Her friends called her Hazy. She was never lonely as a child and could have stayed there forever. But island kids often dream of a future that can only be found on the mainland. Hazel Coombs wanted to be a nurse. In the autumn of 1932, when she was fourteen, she packed her best clothing and left Isle au Haut. A small steamer called *The Daydream* was visiting the island by then, but it was reserved for the wealthy rusticators coming to Point Lookout. Hazel rode on the mailboat that came from Stonington. Her uncle was waiting there to drive her down to Portland where she would attend high school. He lived in that city himself and had no children. But he also had no room for his niece. She was promised a bedroom of her own by a well-off Portland family. In return, she would act as nanny to their children when not attending classes. She ended up sleeping in a small cranny off the kitchen with a curtain as her bedroom door.

Going back to Isle au Haut was not an option. Changes had been happening on the island. By the time she graduated from Portland High School in 1935, there were only seventy-five permanent residents left. Hazel packed her diploma into a suitcase, said goodbye to the children she had cared for, and moved up to Bath to live with her sister and attend nursing school. It was her dream come true and it changed the rest of her life. It was there that she met and fell in love with Phillip Tardiff, a welder at Bath Iron Works. He had been admitted to the hospital with pneumonia and was so ill the nurses checked on him often to see if he was still alive. Hazel married him on January 4, 1937.

At five o'clock Hazel took plates down from the cupboard. As she called for her girls to set the supper table, an announcement came over the radio. Word had just arrived that a plow had broken open the tote road to the logging camps that morning, up in the thick woods of Aroostook County. The workers and their animals had been liberated. More automobiles had arrived to help in the rescue and now the procession was almost to the town of Patten. Car horns were tooting, woolen caps were being tossed into the air, and local housewives were dishing up the warm food they had been cooking all day to welcome the haggard ensemble.

"They must be so happy," said Phyllis, age fourteen, as she laid out the plates. "There were children at the camp, too."

"And animals," said Mary Lou. "I wonder if they had any goats."

Hazel smiled as she took the beans from the oven and slid in the baking sheet of biscuits. A year earlier Mary Lou, now twelve, had rescued a pregnant goat from a barbed wire fence where it had become entangled. When the goat had its babies, Mary Lou found them homes on nearby farms.

"Call your father and brother to come wash up," Hazel said. She put a bowl of homemade pickle relish on the table, next to the butter dish. Out on the Kennebec River the afternoon skies blended into gray water. Nights came early in February. When Phil and ten-year-old David came in from splitting firewood in the barn, David stood warming his hands over the register. There was nothing like a hardwood fire to take the winter chill off one's bones. On cold mornings, when the children crawled out of bed for school, they came first to the kitchen and stood over the register's heat.

Hazel turned off the radio. If she were given one wish, it would be that the baby not come during the night. The drive into town and the hospital worried her. Their nearest neighbors were a quarter mile away. Varney Mill Road was not paved and could be dangerous during the winter months. If rain or wet snow fell on the gravel surface and froze, it became a road made of glass. She could spend a few days with her sister in town as she waited for the baby. It would make Dr. Hamilton happy. Even her parents, May and George Coombs, had finally said goodbye to Isle au Haut and moved to Bath. But Hazel's place was with Phil and the children, and so far the weather was working in her favor. As the Tardiff family settled into chairs around the kitchen table, Hazel used a pot holder to lift the sheet of hot biscuits from the oven.

JAMES HAIGH, OF PORTSMOUTH

In Portsmouth, New Hampshire, a busy seafood wholesaler named James Haigh made plans to leave after midnight and drive his truck north to Maine where he would meet up with one of his lobster suppliers. At thirty-eight and after years of hard work, Haigh was now self-employed. He had been a loom fixer in the local woolen mill when he and Eleanor Jackson first married. But he missed the smells and sounds of the ocean. Now he was doing what he loved, and his company was thriving.

Jimmy often stored his lobsters in "cars," wooden crates with slats, in the back channel of the nearby Piscataqua, a twelve-mile tidal river that marked the boundary between Maine and New Hampshire. Those holding boxes kept them alive until he delivered them to local restaurants and grocery retailers such as A&P supermarkets. He and Ellie were running the James B. Haigh Lobster Co. mostly in the kitchen. A large cooker of lobsters would be dumped into the open sink. Even their nine-year-old daughter, Barbara Ann, helped. Jimmy taught her how to remove the meat from the shell without breaking it. That's how his clients wanted it. The A&P chains bought their lobster in cardboard boxes with plastic lids so that that meat inside was clearly visible.

In the evenings Jimmy Haigh did the bookkeeping, tending to invoices and receipts on the same desk he had built as a high school student. The business was doing well enough that he had hired Earle Sanders to drive the company truck, a red Ford with a boxed-in back. Earle often came home with Jimmy for lunch and was soon like a member of the family. He made truck deliveries to customers and picked up lobsters from their suppliers. Jimmy used his own station wagon, a Ford woody, to make closer deliveries to clients over in Concord. The company expanded when he established contacts with fishermen up in Maine, a state famous for its quality of lobster. Now another load was waiting for him to pick up the next day for Monday's market.

Haigh decided to drive the truck up to Maine himself. He had plans to meet Harland Davis before dawn. It was one hundred and forty miles north to Thomaston. There he would turn south for another ten miles to Port Clyde where Davis would meet him. They would be in Harland's thirty-foot boat and on the way to Monhegan Island before six o'clock. It wasn't the best time of year for trapping lobsters, but Davis had been in touch with fishermen on the island. They were known for lobstering in winter. Their combined catches would fill fifty crates, each one weighing up to a hundred pounds. There was also a wharf with a mechanical hoist at Port Clyde. If all went well, they would load the filled crates from the boat into Jimmy's insulated truck. He could pay Davis and be on the road back to New Hampshire that afternoon.

HARLAND DAVIS, OF PLEASANT POINT

Harland had known since he was a kid that he wanted to be a lobster fisherman. Like many boys in coastal towns, he was born to several generations of fishermen and sailors. He had grown up with half-round lobster traps, wooden crates, trap lines, and bait barrels. His father was a lobster fisherman. His late grandfather had been a smack man, visiting coastal communities and offshore islands to buy lobsters that he would then sell on the wharf. His great-great-great-grandfather had been a sea captain, and *his* grandfather had been lost at sea in Indonesia in 1790. Fishing and the sea were in Harland's blood.

With blondish brown hair and dark blue eyes, Harland Davis had been a popular student at Thomaston High School. He was a member of the glee club for four years, the athletic association for three years, class vice president his junior year, and finally president and class valedictorian his senior year. He did well in public speaking and even had a role in a school play. But there were two things on his mind during that last year of high school: a pretty girl named Mary and lobster fishing. The 1941 *Sea Breeze* yearbook had teased him about Mary. "There's a girl in our class, Mary McLain is her name. Being Harland's 'sweet heart' she has gained such fame."

He had bought Mary a cedar hope chest from a furniture company in nearby Rockland, a beautiful work of art. A hope chest meant they'd be married one day. But that was a decade earlier. A lot of changes had occurred since high school. As often happens with first loves, he and Mary broke up and went their separate ways. On New Year's Eve in 1942, a year and a half after he graduated high school, Harland married Ethel Stebbins, a free-spirited girl who was also her class's valedictorian. When Harland and Ethel divorced after a few years, he married Alice French Church in a double ceremony with his sister and her fiancé. It was late December in 1950. They went to New York City on their honeymoon and bought tickets to *Guys and Dolls*, which had opened a month earlier. Blonde and vivacious, Alice was eight years younger and the mother of a seven-year-old daughter. Now Harland had the added responsibility of being a good father. It was a duty he relished.

The other passion Harland had professed in his senior yearbook came true. He was now a full-time lobster fisherman. In the morning, a buyer from Portsmouth would arrive in Port Clyde before dawn. He had telephoned earlier to say he'd be waiting on the wharf. Harland would pick him up there and they would be on the way to Monhegan Island before the sun rose.

February was not the best time of year for lobster fishing. Many lobstermen lay low from December to April, mending their lines and traps and repainting their buoys. It was not the best time of year for good weather on the water, either. The latest forecast had predicted a storm that would bypass southern and central Maine except for a few inches of snow turning to rain. The ocean might be rough in patches with some minor gusts between Port Clyde and Monhegan Island. But Harland was a seasoned fisherman. Now thirty years old, he had spent much of his life in a boat on the open water. A little bit of wet snow was nothing to worry about.

BROWNVILLE

Fitting perfectly into the center of Maine, the town of Brownville is located a few miles east of what hikers of the Appalachian Trail call the Hundred-Mile Wilderness. It's the longest and most demanding stretch of the entire trip. Brownville grew over the years on gristmills, clapboard mills, and sawmills, all producing a variety of wood products including carriages, shovel handles, matches, and wooden forks, spoons, and knives for picnics. Wood was the needed ingredient to sustain any of these mills and Maine's forests were rich with timber. But it wasn't just wood that built Brownville. That geological area of Maine was also rich in bedrock. The town matured thanks to the impressive slabs of slate and blocks of granite taken from nearby quarries. Iron ore had also been discovered and kept Brownville buzzing for fifty years until it became too expensive to ship. But the wood, the granite, and the iron ore had brought the important railroads to town.[4]

After World War II, the big steam engines that powered locomotives were slowly being replaced across the country by diesel engines, their steam whistles replaced with air horns. But steam still operated most trains that ran through Maine in the winter of 1952, belching rolls of smoke as they chugged along. These trains and tracks that brought business to Brownville also brought a social structure that fell into place over the years. The engineers used whistles—a mixture of long and short blasts—to communicate with other trains or railway workers in the yard. Whistles blew a warning at every bridge and road crossing as the trains rolled past. The children of Brownville were born and raised to the sound of trains. It was the poetry of the railways. It was the language of its workers, and of the people who welcomed the trains into their towns.

SONNY POMELOW, OF BROWNVILLE

Fifteen-year-old Raymond Pomelow Jr., known as "Sonny" to his family and friends, had begun his freshman year at Brownville Junction High the autumn before. His sister, Louise, was in the last class to graduate from Brownville High before it closed its doors a decade earlier. Now Ray and other high school kids in town attended the impressive school that had been built in nearby Brownville Junction, where the railroads intersected. While many folks in Brownville made good salaries in the 1940s and 1950s working for Great Northern Paper Company, the Canadian Pacific Railway, or the Bangor and Aroostook Railroad, there were others who labored as woodsmen. Or they were employed by the mill that made wooden pegs or shoe shanks for the military.

The Pomelow family was considered among the economically disadvantaged in town. Ray Sr. was known to drink more than the typical man who worked hard on weekdays and then let loose on a Saturday night. Sonny's mother, Grace, was often gone from home to cook in the lumber camps, especially during the winter months. It was a way for her to help support the family. As a result, Sonny was mostly raised by his sister, Louise. She saw to it that he had clean clothes, food to eat, and attended school. Nine years older and his only sibling, it helped that Louise loved the boy. Even before she graduated from high school Louise was a hard worker, putting in what hours she could at the peg mill. When she married in 1946, Sonny spent a good deal of time at her house. Louise was a positive influence on her little brother.

Born on June 16, 1936, Sonny Pomelow wasn't a good student. He was a soft-spoken and well-liked boy with average looks. But he was the kind of pupil teachers overlooked in the 1950s so that they could concentrate on the smarter kids. Sonny might have preferred being unnoticed in class. A lot of boys who grew up in mill families saw their own futures there, too. He was more the rule for his economic background than he was the exception. Some of his classmates did pick on him for being slow, as they called it. But his friends accepted Sonny as he was. Some of them even looked out for him.

Despite these setbacks, there had been good childhood memories made on Spring Street where Sonny Pomelow grew up. His house sat next to the high school Louise had attended until it became an elementary school for Brownville. This was the Stickney Hill area also known as Skunk Hollow. The hills on each side of the hollow were owned by a neighbor who didn't mind when local kids came there to try out their Christmas sleds or toboggans. When sun and wind melted the fresh snow and it froze during a cold spell, it left a hill's surface hard as crust. The kids flew like rockets down the slopes, sitting atop pieces of cardboard or chunks of discarded linoleum flooring. When the snow was powdery and fresh, they skied down. And then there was always Frog Pond, at the foot of Stickney Hill. It was a perfect place to skate under the stars on a winter's evening. Or the coves on the Pleasant River until the eight o'clock whistle of the B&A night train was the curfew knell that called them home on school nights.

As with many other American boys, the Boy Scouts might have saved Sonny Pomelow from withdrawing into himself. Shy unless with his trusted friends, he thrived with his local troop. They did community service as they earned their badges and took part in parades and funerals. Now and then they held a musical or a play at the grange hall. Two summers before, fifteen scouts in

Troop 111 went camping for four days on Mount Katahdin to complete their emergency training program, which included what to do in case of a nuclear attack.

There was no favorite girl Sonny had his eye on, though there were plenty of pretty ones in his class. Even if he had been popular, as some of the best athletes at Brownville Junction High were, girls weren't his priority at age fifteen. Cars were. He especially admired the ones in *Hot Rod*, a monthly magazine published out in California. Sonny and his best friends, Johnny Ekholm and Bobby Williams, would stop by the Rexall drugstore in Brownville Junction to buy the newest issue. The boys were devoted fans of hot-rodding. Girls couldn't compete with hot rods, at least not yet.

Sometimes they'd sit at the soda fountain and order a sundae or a milkshake if they had the money. Or they'd crowd into the red Naugahyde booths with the tabletop jukebox that played a song for a nickel, six for a quarter. Johnny would turn the pages of the magazine slowly so they could study each picture. They loved it when cartoon character Stroker McGurk, a futuristic-thinking hot-rodder created by cartoonist Tom Medley, drove his 1929 Ford Roadster across the pages. The boys had plans for when they graduated from high school three years down the road. Bobby had even looked up the distance in an atlas. It was over three thousand miles from Maine to California, where the coolest cars and the drag races were.

"If we work summers and pool our money," Johnny would remind them as he turned pages, "we can do it."

BAR HARBOR, ON MOUNT DESERT ISLAND

In 1604, during the same voyage that brought Champlain within sight of Isle au Haut, the *high island*, his ship sailed on north along the Maine coastline. He was acting as navigator for the French nobleman Pierre Dugua de Monts, who had been sent to North America by King Henry IV to "establish the name, power, and authority of the King of France," as well as "summon the natives to a knowledge of the Christian religion." And they may as well look for precious metals in the meantime. Leaving the principal expedition to explore the Maine coast in a smaller boat, Champlain cruised past the many islands, coves, and reefs. While the land he saw from the water was forested with pine and firs, he was impressed by the towering cliffs with summits void of trees. He named it *L'Île des Monts Déserts*: Island of barren mountains.

Three hundred years later, Mount Desert Island was firmly established and had for decades been home to those "rusticators" who would later discover Isle au Haut. The town of Bar Harbor was overrun with mansions filled with Vanderbilts, Fords, Carnegies, Astors, and Morgans. Opulent parties thrown in parlors and aboard yachts commanded their social calendars. Among the wealthy summer residents were Ernesto Fabbri and his brother, Alessandro, both born and raised in Manhattan with moneyed family connections in Italy. Ernesto, who was described as a linguist and world traveler, had married Edith Shepard, the great-granddaughter of Cornelius Vanderbilt. As 1900 brought the country into a new century the couple was busy building Buonriposo, a cottage in Bar Harbor that could claim status as an Italian villa. This was on Eden Street, five miles north of the towering rock promontory named Otter Cliffs that jutted out over the Atlantic Ocean and had once inspired Champlain.

The unmarried Alessandro spent his summers at Buonriposo with his brother and Edith. More scientist than playboy, his passion was wireless telegraphy, the transmission of telegraph signals using radio waves. And there was no better place for radio transmission than at Otter Cliffs. When Great Britain declared war on Germany on August 4, 1914, gossip soon made its way into Maine newspapers that German spies were afoot in the state. With those invisible radio waves in the skies over Bar Harbor, suspicion soon fell on the Fabbri brothers, despite their attempts to quell what they saw as a ridiculous accusation.

When the United States entered the war on April 6, 1917, Alessandro Fabbri saw it as a chance to prove his loyalty to the country of his birth. He bought the land overlooking the ocean at Otter Cliffs and had it cleared at his own expense. He then directed the installation of a superb wireless telegraph station and offered it to the United States government as part of the war effort. In exchange, he wanted a commission in the Naval Reserve and to be made officer in charge of the station. On August 28, 1917, the Otter Cliffs Radio Station went to work receiving and transmitting signals under the charge of then thirty-nine-year-old Ensign Alessandro Fabbri. It was the best location on the East Coast, free of man-made noise and sitting atop an open span of ocean between Maine and the shores of Europe.

ROCKEFELLER'S VISION AT WINTER HARBOR

While the navy considered the Otter Cliffs Radio Station "the most important and most efficient in the world," the place fell into disrepair after the war ended. Alessandro Fabbri had died in 1922, leaving his vision to weather the coastal storms and harsh Maine winters. By the early 1930s, it had become an eyesore. Driving past the dilapidated sight was unpleasant enough for local residents, but it was more distasteful for the summering millionaires and billionaires. Since the navy had no funds to restore the station, they made a deal with John D. Rockefeller Jr., the financier and philanthropist. Rockefeller already wanted to expand the motor road system on Mount Desert Island. If he agreed to build a new receiving station within a fifty-mile radius, the navy would allow him to widen the road. And the Otter Cliffs site would be donated to Acadia National Park, which Rockefeller was also instrumental in developing. A deal was made.

Five miles across Frenchman Bay from Otter Cliffs was Big Moose Island, at the tip of the Schoodic Peninsula and near the town of Winter Harbor. Rockefeller began plans to build an impressive and expensive complex there that echoed the architecture at his own Maine mansion. The principal building would be château-esque and designed by Grosvenor Atterbury, whose early work included weekend mansions for the wealthy, as well as the elegantly

designed Buonriposo. By September 1935 construction was finished and a
new navy radio station was officially back in business.

In 1950, the station's name became the U.S. Naval Radio Station at Winter
Harbor. Twin radio towers, steel structures 210 feet high, towered above the
grounds. The term "Cold War" was now familiar to American citizens. This
new base reported bearings taken from aircraft, surface ships, and transmis-
sions from submarines to the Atlantic Division control center for the Office
of Naval Intelligence. With its elaborate structures and landscaped grounds,
the Winter Harbor station was a dream come true for any young man lucky
enough to be assigned there.

PAUL VINCENT DELANEY, SEAMAN 1ST CLASS

In late 1951, a handsome young sailor from Staten Island ended up stationed at
the naval base. Joining the navy had not been Seaman 1st Class Paul Delaney's
idea. It was firmly suggested to him by his parents, his father in particular,
that perhaps a stint in the military would give the boy some focus and teach
him a bit of discipline in the process. It wasn't that Paul fell into the scofflaw
class. But he had a sense of adventure that his parents didn't always approve.
The Delaneys were strict Catholics who had survived the Great Depression and
learned the hard way to be cautious. While Paul had done well in grammar
school, his high school years presented a challenge. He wanted to see a world
that he knew existed beyond the Staten Island Ferry. Ignoring their classes,
he and two buddies lit out for the Mardi Gras in New Orleans. When they
returned two weeks later, he was summoned to the principal's office and
expelled.

Having lived through those dark years after the stock market crashed, it
was little wonder that the elder Delaneys saw the future differently than their
children. With no prospects now in sight for their son, they sat him down one
night and suggested he turn his life around before it was too late. His father
thought the military might be the answer. Paul V. Delaney enlisted in the

navy on October 10, 1950, becoming a Seaman 1st Class, the entry-level rank. He had turned nineteen years old three weeks earlier. He was soon enrolled at Class A Radioman School in San Diego for sixteen weeks.

When the good-looking New Yorker with his dark wavy hair and hazel eyes turned up at the U.S. Naval Radio Station at Winter Harbor, he seemed a good candidate for the commander's prediction about the young men assigned there. "You will do one of three things while you are here," the commander told incoming sailors. "Marry one of the local girls, have a car accident, or reenlist." Paul Delaney looked around at the fancy digs and knew he'd make out fine. He had no intention of a career in the navy, or any other branch of the military. He'd already had a car accident serious enough that the police had come to inform his parents and scare the daylights out of his little sister. And at twenty years old, the last thing on his mind was marriage.

But the local girls were certainly pretty. Paul had met one from Bar Harbor, a shy girl with shiny brown hair. Her name was Mona and he liked the way she smiled. The next day was a Sunday and Paul was off duty for a couple days. So he asked his roommate if he could borrow his car.

"I'll fill it with gas before I bring it back," Paul promised.

"Just don't wreck it," Jeffrey said, and tossed him the keys.

SATURDAY NIGHT

As night drew on, and, from the crest
Of wooded knolls that ridged the west,
The sun, a snow-blown traveller, sank
From sight beneath the smothering bank,
We piled, with care, our nightly stack
Of wood against the chimney-back,—

Shut in from all the world without,
We sat the clean-winged hearth about,
Content to let the north-wind roar
In baffled rage at pane and door . . .

—"Snow-Bound: A Winter Idyl"
by John Greenleaf Whittier

JAMES MORRILL AND PETER GODLEY

It was a busy night at Cap Morrill's tavern in Brewer on the outskirts of Bangor. Across the street sprawled the mill owned by the Eastern Paper Corporation. The company had received word a few hours earlier that the lumberjacks cutting timber at their camp up north were no longer marooned in drifts of snow. They were following the Diamond Match Company workers out of the woods to safety. The tavern had steady customers in the local mill workers, beginning at daybreak. Many men stopped in when the night shift ended to have a beer or two before heading home to sleep. The atmosphere in the tavern that night was one of celebration. Behind the bar as usual, Jimmy Morrill kept the drinks coming. He was used to lively Saturday nights. But he had plans to shake off the work week the way he usually did during the winters. He and first cousin Pete Godley had their ice fishing gear packed. They would drive to Branch Lake at daybreak.

Both thirty years old and still bachelors, Jimmy and Pete would throw together some sandwiches in the morning. It used to be that Jimmy's mother, the English-born Annie Burnett, packed the men a lunch worthy of a work crew when they headed out to fish or go boating. But Annie had passed away the previous year. She and her sister Marion had first visited America in 1912. Their Scottish father, a papermaker, had already emigrated to Brewer because of the large mill there. On that visit to Maine, Annie met Henry Morrill and fell in love. They raised their family in Brewer. Marion returned to England and married John Godley. Their son Peter was born and raised there.

Henry Morrill was already in business when he met Annie Burnett. He had opened a tobacco and cigar business in a three-story building on South Main, across the street from the mill where he had once been employed as a machine tender. Over the years, it grew into a successful drinking establishment. Eventually he and Annie had four sons, Henry Jr., Carleton, Richard, and James, all serving in the military during World War II and all returning safely to Maine. When it came time to expand on the tavern, the sons had gone into the Maine woods, felled the timber, and then milled it themselves. They called the place Cap Morrill's, since Henry Sr. had been nicknamed "Cap." When he died in 1947, he left the tavern to his sons.[5]

The Morrills were a close family, both in business and socially. They lived in a three-story house, the original building for Cap Morrill's, next door to the present tavern. Jimmy and cousin Peter shared the first floor with Annie until she passed away. Brothers Carleton and Richard—Henry Jr. had left the tavern business to open a service station up the street—lived with their wives on the two upper floors. Hard workers, the Morrill brothers also cut and sold Christmas trees each season. They shared holiday celebrations and family outings. If one brother needed a hand, another brother was just a shout away. When their cousin Peter Godley had arrived from England, he fit into that tight family structure as though he'd been born to it.

Like the Morrill sons, Peter Godley was also a seasoned military veteran. Before he emigrated to the United States, he had served in the British Eighth Army. He ended up in North Africa with the Coldstream Guards, being chased across the desert by Rommel. Many of his comrades didn't survive. Winter or summer, Pete rarely missed an outing with Jimmy. He had learned that nature brought him a measure of inner peace.

Jimmy had been bartending at Cap Morrill's since his discharge from the military. These trips had a deeper meaning for him. There were times out on the expanse of lake, surrounded by conifers and blue sky, his lines in the water, that he remembered days when he thought he'd never see Maine again. His regiment had stormed ashore at Utah Beach on June 6, 1944, one of five designated beach landings in northern France. It would become known to the world as D-Day, the largest invasion by sea in history. Both Jimmy and Peter had survived the horrors of World War II. It was a spiritual bond that sealed their friendship, though they rarely spoke of it.

Branch Lake was one of their favorite fishing spots, winter and summer. It was close, too, just a twenty-five-mile drive southeast on Route 1A. The weather forecast for the next day had predicted a few inches of snow in the afternoon, turning to rain. Since they would be taking Peter's car, a British-made Hillman Minx, a couple inches of snow was no reason for concern. The Minx was like riding in an armored truck. They would bring Laddie, the German shepherd owned by brother Richard but whose affections were shared by everyone. In the morning, when they packed sandwiches, bottles of pop, and homemade pastries, Jimmy would throw in a few raw hot dogs for Laddie.

THE CITY OF BATH

It was a typical Saturday night. Goddard's Pond at the south end of town was packed with skaters, their voices ringing in the cold air. There had been a full moon a few days earlier and now a waning moon hung in a sky filled with stars. Downtown on Front Street, the huge face of Hallet's clock kept an eye on customers strolling in and out of Hallet's Drug Store, as it had for seventy years. The Bath Opera House, with its elegant stained-glass doors, was offering the film *Hong Kong*, starring Ronald Reagan and Rhonda Fleming. Moviegoers at the Uptown Theater had just come from seeing cowboy star Whip Wilson in *Cherokee Uprising*.

As usual, Sam's restaurant on the corner of Middle and Pine streets was packed with diners, most ordering the celebrated rolls made from fresh lobster caught in nearby waters. Sam's French fries—they were cut from dirt-covered potatoes and thrown unwashed into the hot grease—were professed by locals to be the world's best. The bar over at the historic Sedgwick Hotel had a busy crowd. Because the hotel sat across the street from the courthouse, many law cases had been settled there over cocktails. This night there was excited talk of the chartered bus that would leave Bath the next day to transport fans to an Ice Follies matinee at the Boston Garden. A half mile away on Water Street, the rowdy bar at the American House Hotel, with its shipyard clientele, was rocking. The few streetwalkers near that establishment were as relieved to get out of the cold wind as they were to be paid for their services.

A weather prediction from WPOR that a northeaster would pass over the next day with a bit of snow and rain was so routine it went unnoticed.

WILLIAM DWYER AND SNOOKY THE CAT

Sixty-seven-year-old Bill Dwyer opened his front door and stared out at snow from the last storm glistening beneath the porch light. He had been a widower ever since Nellie, his second wife, passed away two decades earlier. Bill expected they would spend their last years together, but Nellie died suddenly at the age of forty-nine. His loyal companion now was a yellow cat named Snooky, with matching yellow eyes. Snooky had gone out two hours earlier for his nightly neighborhood prowl. The cat's paw prints, leading down the porch steps and out into the yard, were still evident in the snow. Bill could usually set his clock by Snooky's return. After a few bites from a bowl near the kitchen stove, the cat would then curl on the sofa to snooze.

Born, raised, and schooled right there in Bath, Dwyer had held a lot of jobs over the years, all local and all marine-related. Mostly, he had been a riveter in the shipyard at Bath Iron Works. But his passion was his connection to the fire department. Until a few years earlier he had been a trained volunteer for Engine 4, a *call man* since 1918, stepping in for emergencies. It was during his time with Engine 4 that a call came in for a cat rescue at Cyr's Market. Bill remembered it was a cold, drizzling day at the end of February, two decades earlier. When the firetruck pulled up to the curb, Mr. Cyr had pointed to the roof of Greenblatt's Tailor Shop next door. There crouched a terrified calico cat that seemed ready to jump.

"That's Snooky," Mrs. Cyr informed the firemen. "She's expecting kittens any day."

A cat lover, Bill knew he would be the one to climb the ladder. He hoped a terrified feline wouldn't fly into his face when he reached for her. Or leap twenty-five feet to the street below, not a good idea in her condition. But when he lifted her from the roof, the cat snuggled against his chest. Cheers went up from the market customers watching down below. The local newspaper printed the story, having a bit of fun with a cat's reputation for possessing nine lives. "Firemen from Engine 4 are probably responsible for the saving of 45 to 63 lives late Friday afternoon." Over a decade later, when Bill began feeding a scrawny yellow stray that turned up at his door, he decided to keep it. Nellie would have wanted him to. "You hungry, Snooky?" he asked, as the cat rubbed against his leg. He had always liked that name.

"Come on, yellow cat," Bill called now. He waited for an answering meow. Instead, the sharp sounds of metal against metal at Bath Iron Works drifted over from where the shipyard sprawled along the riverbank. The third shift would come on at midnight, paid double for losing their weekend. Bill knew the schedule well. His riveting job had kept him there for two decades. But he was a younger man then. Riveting was a demanding, ear-splitting job, but one the crew took pride in. It was an art form now being lost as more and more riveters were replaced with welders. It wasn't just about money, but speed. Welding was faster and saved time. Bill closed his front door against the nightly chill. He wasn't worried about the cat. Snooky sometimes met up with unexpected excitement down the street and forgot his curfew. He'd be home soon.

THE TARDIFFS, ON VARNEY MILL ROAD

Hazel Tardiff was working on a dark blue mitten she was knitting for Mary Lou. She had lost count of the mittens already knitted that winter. It seemed that the kids were forever losing one at school or forgetting one on the bus. She hoped to finish this pair before the baby came. The phone rang and Phil answered. It was her mother, May Coombs, calling to ask how Hazel was feeling.

"I might turn up at your door on Monday with my suitcase," said Hazel. "Dr. Hamilton thinks it's best for me to stay in town until the baby comes."

May agreed with the doctor. Sometimes, it seemed to Hazel as if the lives they had all led back on Isle au Haut—the lighthouse, the cormorants at Boom Beach, the sun over Coombs Mountain—were just dream lives. It was all so different on the mainland, with her father now working in the carpentry shop at Bath Iron Works. But with the good things about island life came the bad things.

May Jarney had come to Maine from San Francisco as a sixteen-year-old orphan. She had met and married George Coombs on the mainland in 1907. George then moved his bride to Isle au Haut where they could raise a family. Hazel was their third child. Three years before she was born, her mother gave birth to a stillborn girl. This was a few days before Christmas in 1914. May had borne two children by this time, six-year-old Evelyn and a two-year-old boy. Old enough to remember, Evelyn had listened to her father's hammer, nail after nail, as he made the baby's coffin. She had told Hazel about the muffled sobs coming from May's bedroom as George left with the coffin on his shoulder. Evelyn watched through the window as her father carried the small box down the snow-covered road to Coombs Cemetery. The burial could happen only after the spring thaw. On an island, or on the mainland, things could go wrong with a baby coming.

"We'll expect you on Monday then," May said. Hazel had never mentioned that lost baby to her mother. Sometimes, it's best to let old sorrows lie.

Hazel returned to her knitting. A radio played upstairs and she could hear her girls laughing. David was in his own room catching up on homework for Monday. In a few minutes, before he went to fill the furnace with wood, Phil would call up to them that it was time for lights out. Feeling the baby move now, Hazel placed a hand on her stomach. Phil saw this and smiled. He folded the newspaper he'd been reading and lay it on the coffee table. His stubbed his cigarette in the ashtray.

"It will soon be over," he said. "Everything will be fine."

Hazel nodded. She had promised to telephone Virginia Hamilton the minute she felt the first labor pain. The doctor's number was scrawled on a piece of paper in a kitchen drawer near the wall phone: DR. HAMILTON H13-5171.

"Maybe I won't take the kids to church in the morning," Phil said.

He had gone with the children the week before while Hazel stayed at home. Sunday worship posed an added problem in winter. St. Mary's Catholic Church was still heated by coal. It was built in 1856 as a replacement for the Old South Church that had been burned by the Angel Gabriel's angry mob of anti-Catholics. St. Mary's had endured and was now a beloved landmark. But even the most devout believers had learned to bundle up well during the winter months. And then, the state of Maine had been hit with an epidemic of measles. Just that week the newspapers listed Bath as having the second highest number of cases. If one of the Tardiff children became infected, it was likely the other two would as well. And Hazel was coming home soon with a newborn baby.

"We're supposed to get an inch or two of snow tomorrow," Phil added.

That it was several miles to town was always a consideration. But Hazel knew her husband was reluctant to leave her alone, even for a couple of hours. Considering the weather and the measles, a Sunday at home was a good idea. On Monday, she would pack a few things and go stay with her parents.

"I doubt the kids will be disappointed over missing church," Hazel said.

SONNY POMELOW, IN BROWNVILLE

The evening whistle of the B&A train had already called the kids home from skating on the Pleasant River. Sonny Pomelow, skates slung over his shoulder, walked home with his friends. Overhead, the Big Dipper glittered in the night sky. Johnny Ekholm pointed at a bright cluster of stars.

"Orion," he said. "There's his belt."

"It's cold tonight," said Sonny. Even wearing gloves, the tips of his fingers tingled. He shoved his hands into his jacket pockets.

"It's gonna drop below freezing," Bobby Williams agreed. "But that's a lot better than below zero." Some winters they had seen the wind chill factor take the temperature down to forty below. But so far, as the weather pundits had predicted, it was a mild winter for Maine.

"It's always warm out in California," Johnny reminded them. "If everything goes as we planned, that's where we'll be in three years."

Sonny couldn't imagine a place where it was warm all the time. There would be no heavy coats to button up, or mittens and gloves to keep their hands warm. But there would also be no sliding down the hills of Skunk Hollow, or skating on the river, or snowball fights from behind snow forts they had built themselves. And what must Christmas be like if everywhere you looked you saw palm trees? Maybe he would get used to it. But then, they planned to be gone for just one summer. The Maine winter would wait for them to finish their adventure.

"You guys doing anything tomorrow?" Bobby asked.

"If it don't snow too hard, we're driving down to Gordon's birthday party," said Sonny. Gordon Joslyn was his five-year-old nephew, sister Louise's son. She and her husband had moved fifty miles south to find work. They were now living west of Bangor in the tiny town of Etna.

"If we get a lot, I have to shovel our yard," said Johnny. "Dad already said so."

Three inches of snow had been predicted for the Brownville area as the newest storm passed over them, on its way up to Aroostook County. But not trusting weather reports was almost a ritual among rural Mainers. Three or four inches of snow could be removed by the plow trucks as fast as it fell. But Sonny knew that if it snowed harder than that, the birthday trip would be canceled. The roads weren't great down to Etna as it was. He would also have to get out a shovel. Most boys his age were expected to clean driveways and porches. Many of them got paid for it. Sometimes, the neighbors hired Sonny. It was a hard job, but it was a way to save money for California.

"See you guys later," Sonny said. His words hung frozen in the air.

Before he turned and headed home to Stickney Hill, Sonny Pomelow stood looking up at the heaven of stars that blanketed Brownville. At times like that, when he was alone with his thoughts and the stars sparkled down, he felt like maybe God was talking to him. It was a mysterious language, as if the stars knew something he didn't. Or maybe his Grandfather Pomelow was sending him a message. A farmer once, the old man had passed away just three weeks earlier in a Milo nursing home. Eighty-five years old, John Pomelow had been sick for a long time. Sonny was too young to know his grandfather well, but he was one of four grandsons who were pallbearers. He didn't care for funerals, with the sad faces and muffled sobs of family members. Even the scent of all those flowers reminded him of a hospital. But Sonny had put on his best suit and helped carry the casket in memory of his grandfather.

A Chevy pickup passed on the road and tooted. It was Donald Stickney, the man for whom Stickney Hill was named. Sonny waved his arm in recognition, hoping Mr. Stickney would stop. He liked pickup trucks, but he dreamed of owning a fancy car when he was older. Maybe even a Henry J. They had been in production for about a year but Sonny had yet to see one that wasn't in a magazine. He'd need a better job than shoveling snow to buy a car like that. When the red taillights braked a few yards down the road, Sonny ran toward the truck, thankful for the warm ride home.

THE SEAMAN FIRST CLASS, AT WINTER HARBOR

Paul Delaney had finished his duties. As a seaman first class, he had the messier jobs, like cleaning toilets in the officers' quarters and mopping the floors. Or peeling potatoes and taking care of garbage and trash. In many ways it was like being home again, with his mother giving him errands to do. But he knew he was paying his navy dues. He would move up in rank before his hitch was over.

Being assigned to the naval station in Winter Harbor was an experience in itself. The main structure had been named "the Apartment Building" by the architect. With its French Revival style, it was *H*-shaped in layout, two and a half stories high, with terra-cotta roofs. The main floor had a massive brick fireplace and impressive staircases. The eleven apartment units on the first and second stories were the primary residences for the officers, with built-in medicine cabinets in the bathrooms and fully equipped kitchens. There were also offices and rooms where radio operations were carried out. In the back was an impressive courtyard enclosed by pavilions, and an expansive terrace.

Paul Delaney and the lower-ranked personnel were assigned to Quonset huts that were erected not far from the Apartment Building. They didn't mind. All around them had been planted a variety of trees, including yellow birch and white spruce, and dozens of flowering shrubs, even blueberry bushes. The circular drives were bituminous and the pathways graveled. A tennis court had been built a decade earlier. While officers enjoyed the best perks the place had to offer, the enlisted sailors relished being at Winter Harbor and the status it gave them in the minds of many local Maine girls.

Off-duty for the next two days, Paul would sleep late for a change and then leave for Bar Harbor before noon. He had plans to invite Mona for a bite to eat after the matinee was finished. They could get to know each other over hamburgers and fries at the local diner. He liked her name. It reminded him of Nat King Cole's big hit, "Mona Lisa." He wasn't looking for a permanent girlfriend, not yet. He was still too uncertain of his future. But Mona was so pretty, and she had a smile that made you wonder what she was thinking, just like words to the song. Paul tossed the keys his roommate had given him onto the table next to his cigarette lighter. He'd catch a bit of television in the community room before he turned in.

PORTSMOUTH, NEW HAMPSHIRE

Once he got to Maine, Jimmy Haigh planned to take the new stretch of turn-pike that had just been finished, from Kittery to South Portland, a distance of forty-five miles. He would then exit onto U.S. Route 1. He watched as Eleanor packed sandwiches and donuts into a paper bag for the trip. He would leave Portsmouth around 1:30 A.M. and didn't want to wake her.

"Stay in bed when I leave, Ellie," he said.

But she had never listened before when he made his trips to Concord, or the few up to Maine in order to expand the business. He knew she would be up after midnight and already boiling water to make hot cocoa for the thermos that would ride on the seat next to him. The truck's heater worked well. Otherwise, it would be a cold ride after he left the turnpike and followed the ocean on Route 1.

He caught Eleanor looking at him. She had told him that morning she didn't feel good about this trip. She couldn't explain it, but a dark mist had hung over her all day.

"Jimmy, I'm worried," she said now. She knew her husband could take care of himself. He had a reputation with the local fishermen for knowing the ropes. He once made the New Hampshire papers for catching with a grapple anchor a giant sunfish that had wandered into coastal waters. In a small boat, Haigh had waged a three-hour battle with the half-ton fish. It was the second largest sunfish caught in New England and the twelfth largest ever recorded in the world. Two professors at the University of New Hampshire had preserved it for use in their marine biology classes. It was a story Ernest Hemingway might have written.

"I'll be home by suppertime," Jimmy assured his wife.

When he had caught up with the paperwork on his desk, he went upstairs to kiss nine-year-old Barbara Ann goodnight. He pulled the blankets over her shoulders and stood for a few seconds listening to the sound of her breathing as she slept. They were his girls, Ellie and Barbara Ann. He was a lucky man. He leaned down and kissed his daughter once again.

In their downstairs bedroom, Ellie was brushing her hair and getting ready for bed. She said nothing more about the trip. Jimmy took off his eyeglasses and lay them on the nightstand next to Barbie's school picture. It was time to get some sleep.

PLEASANT POINT, MAINE

The ocean had disappeared into twilight, with just the sound of it now slapping against the wharf. The silver poplar that would boast rustling leaves in the spring stood stark against the water. It was getting late, but Harland Davis had his boat, the *Sea Breeze*, ready for morning. Before going up to the trailer, he did what he often did. He stood staring out at the dark water. The job his grandfather had done, from before the turn of the century up to his retirement days, had been as a smack man. Richard Davis ran a small boat on a regular route to buy lobsters from fishermen who lived on offshore islands or in nearby villages. The boat was equipped with a tank in its hold that had punched holes so that seawater could circulate and keep the lobsters alive. He could carry six or seven hundred pounds of lobsters in that tank. He motored them back to the wharf and sold them there.

But the job as a smack man had been eased out by the late 1930s. Harland now did the same business as his grandfather but with a different method of operation. He was a land-based dealer, a go-between agent for the fishermen and wholesalers. He wasn't just good at his job. He was liked and trusted by those men who made hard livings from the sea. After all, he was one of them.

Harland had made arrangements earlier in the week. He would buy fifty crates of lobsters from Monhegan Island fishermen that he would sell to Haigh, the wholesale dealer from Portsmouth. Everyone made decent money. And tomorrow there would be a lot of lobsters waiting for them on Monhegan. Each crate could hold a hundred pounds, a lot of meat. It would be a profitable day. He left the wharf and walked up the narrow road to his trailer, now silhouetted on the evening sky. He had bought the small mobile home and moved it onto a parcel of land before he and Alice married. There was even a tiny bedroom for his stepdaughter. They could save money for a future house that way, and they would soon need one. Alice was almost four months pregnant.

Before he went inside, Harland stood for a moment listening to the sounds of Pleasant Point at night, of canned laughter coming a rare TV set in a neighbor's house, a lonesome truck shifting gears out on the highway, waves curling against the wharf. Yellow lights in the scattered houses dotted the winter twilight. Across the road from his trailer sat the comfortable farmhouse where his parents still lived. Harland had been born in that house. As a boy he milked cows in the barn next to it and collected eggs from the hen roosts. His life had always been in Pleasant Point. His future would be, too.

Pausing on the rug Alice had placed at the trailer door, he kicked snow from his boots. His fingers stung with cold, and he warmed them over the register's heat. Alice had gone to bed, but she left a plate of food on the stove. When Harland had eaten, he put his dirty dishes in the sink.

CAP MORRILL'S, IN BREWER

Just before eleven o'clock, Cap Morrill's Tavern gave last call. Jimmy Morrill was saying goodnight to his regular customers as he wiped down the bar and pulled the last beer glasses from the bin of dishwater. By sunup he and cousin Pete would be on their way, Laddie excited and bouncing around in the back seat. Pete had purchased the live baitfish earlier in the day. They had already packed their fishing gear into the trunk of his car, the wooden tip-ups with the red flags and the five-foot hand chisel for drilling holes in the ice. Jimmy would boil strong coffee at daybreak to wake himself up. Pete would have his cup of English tea. Then they'd get on the highway. Branch Lake would be theirs for all of Sunday.

BILL AND SNOOKY

When Bill Dwyer opened his door to call Snooky in for the night, no flakes were fluttering down over Bath. Cold snow from the last storm still sparkled peacefully beneath the streetlights. The weathermen had apparently gotten it right and northern Maine would be the recipient of the incoming northeaster. When the yellow cat meowed against his leg and was safely inside, Bill snapped off the porch light and closed the door.

Little villages along the seacoast and inland had heard on their radios the latest weather update from WPOR in Portland, that a heavy storm was estimated to hit Maine the next day, a northeaster. And because nature doesn't think in terms of fairness, it would again concentrate on the northern part of the state. The rest would see a "mild amount" of snow that would turn to rain. Expecting the new storm to pass over them, coastal, central, and southern towns settled in for the night. They filled their furnaces with hardwood or loaded them with coal, then turned back the bedcovers. For many folks there would be church in the morning.

While they slept, the storm moved north toward New England.

PART TWO

FEBRUARY 17, SUNDAY

A hard, dull bitterness of cold,
That checked, mid-vein, the circling race
Of life-blood in the sharpened face,
The coming of the snow-storm told.
The wind blew east: we heard the roar
Of Ocean on his wintry shore . . .

—"Snow-Bound: A Winter Idyl"
by John Greenleaf Whittier

northeaster is a fickle force of nature. It's an extratropical cyclone, so-called because it occurs outside the tropics. These storms can be born at any season of the year, not just during winter. But the strongest and most violent ones typically happen from September through the month of April. All bad weather results because the sun doesn't do a satisfactory job in equally heating the planet. It works best at the equator, which heats more intensely than at the poles. But consider what the sun has to work with. For one thing, the earth is a tilted sphere made up of 70 percent water and 30 percent landmass. The sun's heat penetrates water at a different rate than it penetrates land, which heats more quickly. That the earth doesn't stand still while this is going on but rotates on its axis—half the sphere is in daylight while half is in night—adds to the problem.

This inequality in heating creates air masses of different temperatures, which constantly strive to equalize. This sets the atmosphere in motion: cold air moving southward with warmer air moving northward. When these air masses collide, the colder air displaces the warmer air, causing it to rise. The result is often stormy weather. But occasionally these air masses meet each other and come to a gridlock. When that kind of congestion occurs, something has to develop so the air will move again. That something is an area of low pressure called a cyclonic storm. It rotates in a counterclockwise direction, forcing warmer air northward on one side, and colder air southward on the other. The temperature differences between the air masses, and the upper air wind patterns that force the air to rise vertically, energize a cyclonic storm.

These types of storms often form off the East Coast of the United States, between Georgia and New Jersey, within one hundred miles of the coast. This meeting place is the favorite breeding ground for northeasters. During the winters, the polar jet stream carries frigid Arctic air south across the Canadian plains, eastward to the Atlantic Ocean. There it meets warm air from the Gulf of Mexico that is moving northward. Northeasters flourish on those converging masses of air. They feed on them as if they are fuel. Under certain atmospheric conditions, they can obtain hurricane-force winds. Once nourished, they move northward, often intensifying along the way. As they approach New England, the counterclockwise flow around them results in a northeast wind—provided the storm stays to the east, over the ocean. Northeasters often reach their maximum power near New England or the Maritime Provinces.[6]

The 1952 northeaster that would arrive on the heels of the storm that had incapacitated the lumber camps in northern Maine began its initial formation in the Gulf of Mexico. On February 16, it redeveloped offshore and downstream from Cape Hatteras on North Carolina's Outer Banks, the highest risk area for storms and hurricanes on the Eastern seaboard. The 16th was a Saturday, the same day a caravan of woodsmen, women, and children trekked back to civilization, bringing their haggard horses and cattle with them. This extratropical cyclone would develop rapidly in the next twenty-four hours.

Its center would stay out to sea as the storm moved offshore from southern New England. There it would snatch moisture from the ocean to produce an enormous amount of precipitation. The temperature was cold enough that this precipitation would fall as snow.

Even the milder northeasters can be uncaring and harsh visitors, bringing destructive winds laced with heavy snows or lashing rain. And it's not unusual for a susceptible region of the country to be hit by a storm like this several times in a single winter. A modern classification scale ranks storms in these categories: 1 is *Notable*; 2 is *Significant*; 3 is *Major*; 4 is *Crippling*; and 5 is *Extreme*. Categorizing a storm by using this system is not a fail-safe method. It takes into consideration the number of people impacted in the region affected and the inches of snowfall reported. It does *not* factor in those elements that can dangerously affect the power of a storm: gale forces, blizzard conditions, freezing rain, ice storms, tidal surges, and coastal flooding. This incoming northeaster would be only a Category 1, or *notable* storm. Yet it would bring with it the heaviest snowfall of the winter for eastern New York and southern New England.

As Mainers slept, heavy winds circled the storm center in a counterclockwise spiral. Instead of heading out to sea, the storm turned inland, with New England sitting at ten o'clock. Every respectable blizzard holds in its fist two major decider cards: snow and wind. And the "notable" storm that was approaching Maine on February 17, 1952, had those two elements in spades. Meteorologists might have missed it, but the *Old Farmer's Almanac* had predicted for that day, "Now comes a real blizzard right up to your gizzard." While human beings do not have gizzards, they certainly know when they are overpowered by snow.

HAIGH AND DAVIS: MONHEGAN ISLAND

It was just after five o'clock and still dark when Harland Davis buttoned his burgundy-and-gray barn coat with the fleece lining and left Pleasant Point in his lobster boat. He had reassured Alice when he kissed her goodbye. "Honey, don't worry." That evening, he would have a hot supper with her and Carolyn, the little girl who was already like his own. His plan was to run from buoy to buoy, going in by Hupper Island to the wharf at Port Clyde, two miles distant. If James Haigh had made respectable time on the road, Harland would see the lights of his truck on the dock, and Jimmy likely pacing back and forth to keep warm as he waited. The latest weather report predicted that a storm would pass over coastal Maine. Like any fisherman familiar with the ocean, Harland didn't trust the weather. It could turn on a dime. The sooner they got to Monhegan Island and loaded the lobster crates, the sooner they'd get back to Port Clyde.

The morning was breaking clear. The ocean was calm with small wavelets unfurling in a light wind. He steered the thirty-foot *Sea Breeze* out into the water and then opened the throttle. His maximum speed on a good day was seven or eight knots, or about nine miles an hour. As he crossed the mouth of the Saint George River he could see lights blinking on in distant houses, yellow sparks in the predawn. Fishermen on both sides of the bay were up and about for the day even though it was the Sabbath. Sundays brought a different kind of work, getting the buoys ready, doing needed repairs on the nets. Otherwise, the villages were quiet, the morning stillness broken only by the later sound of church bells. February meant coastal Maine was without the summer people, still unspoiled by cars with out-of-state license plates and the din of tourism.

Behind him, in the boat's wake, was the road Harland had ridden up to Thomaston for four years of high school. A couple miles from his house, curving through South Cushing, it passed Hawthorne Point Road where the colonial farmhouse belonging to the Olson family stood. Its weathered clapboards and a woman named Anna Christina had been captured on canvas three years earlier by a young artist named Andrew Wyeth. It was just another old farmhouse to Harland and the local residents, with too many rooms to heat in the winters. And then the newspapers wrote a story that the painting, called *Christina's World*, was becoming well-known to tourists. Gossip was that most of the neighbors had avoided going into the Olson house, given its squalor and the bad smells. And then Miss Olson was fifty-five years old when she was painted, far from a young girl. The artist had used Christina's withered arms and pink dress, but the model for her head and body was that of his own young wife. It was hard to understand artists and city people sometimes.

Harland could have asked Jimmy to meet him at his own wharf. Just three years old, it was 134 feet long, an impressive sight. He and a few friends built it from locally milled spruce. He was proud of his wharf, a symbol of his business acumen as well as his fishing skills. The writer Elisabeth Ogilvie had published a book a few months earlier that mentioned it, and how Harland had painted the derrick bright orange. But the back roads leading down to Port Clyde were more accessible from Route 1, especially in February. Even more important, Port Clyde was the closest wharf to Monhegan Island. That "mild amount" of snow and rain predicted for coastal Maine was best avoided on the open water. The faster the job got done, the sooner they would be back on the mainland.

Gulls circled the *Sea Breeze* looking for fish, their cries barely heard above the purr of the motor. Port Clyde loomed up ahead. The village, sitting at the tip of the St. George peninsula, had grown up on shipbuilding in the nineteenth century. Years later it became a major center for the catching and canning of seafood, so much so that the harbor was once named Herring Gut. To Harland's right was Hupper Island, just a few hundred yards distant yet hidden in morning mist. Across the water to the southeast the electric beacon from Marshall Point Lighthouse flashed a greeting. In the bell tower was a half-ton bronze bell. When wound in stormy weather its hammer struck a warning every twenty seconds. He had grown up to the sounds of bells and whistling buoys, to foghorns and seabirds, and the rugged ebb and flow of the ocean. This was *Harland's* world.

Jimmy Haigh was already tired. The alarm had gone off shortly after midnight. An hour later he was on the road, rubbing sleep from his eyes and drinking hot chocolate. As he knew would happen, Ellie was awake and in her blue housecoat, making sure he didn't forget the lunch she had packed, and that the thermos was under his arm. He had given Barbie a quick kiss on the forehead, not wanting to wake her. But Ellie held him back at the door, wrapping her arms around his waist.

"Take Earle with you," she said. "You could fall asleep on that narrow road." He had seen the worry in his wife's eyes.

"Sweetheart, go back to bed," he said. "I'll be home before you know it."

When Jimmy turned off the short stretch of turnpike onto Route 1, he knew that Ellie was right. It was a tough road to stay awake on. During winter it was mostly a ribbon of white that curled through the coastal towns, a hypnotizing highway. He arrived in Thomaston by 5:00 A.M. as planned. He then turned the truck south, taking a couple of rural highways another ten miles down to Port Clyde where he would meet Harland. Port Clyde's wharf had a mechanical hoist for loading full crates of lobsters from the boat when they returned.

Jimmy parked the truck and poured a cup of hot chocolate from the thermos. At just over six feet tall, he needed to stretch his legs from the drive. A mild case of polio several years earlier had restricted his movements for a time. Working out at the gym, he had finally recovered. But now and then his leg muscles still felt the old illness. Lights were on in some of the shingled houses built within view of the dock. The general store, just fifty feet away, had a sign on the door: CLOSED SUNDAYS. It was the one day of rest in a week of hard work and many folks welcomed it. Jimmy walked a few feet out onto the pier. Ocean winds ruffled the folds of his coat. He and Harland had talked on the phone the night before about the incoming storm that would mostly pass over them. It was the reason they wanted to be on the water early.

He saw the running lights on Harland's boat before he heard the inboard motor. Sunrise would not be for another fifty minutes. He had brought empty crates with him from New Hampshire. He and Harland quickly loaded them into the *Sea Breeze*. The men had bundled well against the sharp wind, both wearing winter coats and heavy rubber boots. Now in the boat, with Harland steering them out toward the channel, Jimmy felt his energy returning. Those February sea winds could wake a man up. He was eager to get the deal done, and then get the hell home before dark. Like Davis, Jimmy Haigh didn't trust Mother Nature.

Harland Davis could close his eyes and steer the boat. Fishing was a job and it was a damn hard one. Yet he never stepped into the *Sea Breeze* without feeling a surge of pride mixed with freedom. He glanced over at Jimmy, who was staring toward the Marshall Point Lighthouse beacon. Harland planned to run the channel by Marshall Point, out through the channel between Burnt and Allen, two of the Georges Islands. It was in those waters, in May of 1605, that the crew of the *Archangel* had harvested the first lobsters in that part of the world, a place they called "the north part of Virginia." Captain George Weymouth of England had landed the *Archangel* on Monhegan Island. Two days later, off the coast of Georges Islands, crew member and explorer James Rosier, already no stranger to that coastal area and the local rivers, had written in his journal that, "towards night we drew with a small net of twenty fathoms very nigh the shore: we got about thirty very good and great lobsters." A lot had changed in 350 years when it came to lobstering in Maine.

TO MONHEGAN ISLAND

Ever since the Great Colonial Hurricane in the summer of 1635 when a British galleon named the *Angel Gabriel* crashed against the rocks a few miles southwest of Port Clyde, shipwreck sites have dotted coastal Maine. Sloops-of-war, schooners, passenger ships, steamers, brigs, freighters, and fishing vessels have all met various degrees of fate in those waters. Named *Monchiggan*, or *out-to-sea island* by the Algonquins, Monhegan knew tragedies. In June of 1941, as Harland was graduating from high school, an excursion boat to the island carrying thirty-four passengers and crew disappeared into a fogbank and was never seen again. An explosion was believed to have caused it to sink.

The *Sea Breeze* was a few miles past Burnt and Allen islands when the sun broke over the water, an orange ball cresting the horizon before it disappeared in clouds. Harland pointed and Jimmy's eyes followed his direction. Straight ahead lay Monhegan, sea-ringed and rugged. Its lighthouse appeared out of morning mist, perched on a hilltop at the island's center. As they drew closer they noticed activity on the wharf, a scattering of men moving to and fro. Likely the fishermen were waiting for them. It was not yet nine o'clock when Harland edged his boat up to the Monhegan wharf. The lobsters would be stored in a lobster car, a pen submerged below a floating wooden raft. They would be already plugged, a wooden peg inserted in their crusher claw, the larger and stronger of the two with coarser teeth. Otherwise, confined to the crates, the creatures would resort to cannibalism.

"Let's get these out of the boat," Harland said. He and Haigh began lifting empty crates up to the hands above that accepted them. These Monhegan men were Harland's friends. One of them pointed out that the cloudy sky was now on its way to being overcast. Gray skies in the northeast were common during winter. An overcast sky with a heavy cloud blanket could even help warm the earth below by trapping heat from escaping back into the atmosphere. But the sky over Monhegan Island that morning looked ominous. Fishermen know skies. Time was of the essence. Before they stacked the crated lobsters into the *Sea Breeze* for the return trip, Harland and Jimmy would quickly eat the sandwiches they had brought with them.

THE COUSINS IN BREWER

By the time Haigh was driving his truck onto the wharf at Port Clyde, James Morrill and Pete Godley were in Pete's car and headed southeast on Route 1A. Handsome, known for his British reserve, Pete had never hunted or fished until he came to Maine and bonded with his first cousin, Jimmy Morrill. He had been born and raised in the historic village of Pontefract in West Yorkshire.[7] With his unmistakable English accent, and behind the wheel of the Hillman Minx, Peter Godley might have stood out in small-town Maine as a conspicuous foreigner. But the Morrill family had welcomed him so warmly when he arrived at their doorstep after the war that Maine now felt like home to him.

Jimmy Morrill could almost smell the fresh snow on the lake. It was a memory of being in the midst of nature that had helped him survive the bloody fighting and the useless deaths he had witnessed during the war. It had been this same time of morning, 6:30 sharp, that he and his fellow soldiers had waded through waist-deep sea while enemy gunfire chopped the water around them. The code name was Utah Beach, and the objective was to secure a beachhead on the peninsula. *Home* was the magic word that had kept him going. Of the four sons in that family, he was the only one not married. Pete was also single, although he now had his eye on a pretty girl at the A. J. Tucker Shoe Factory where he worked. It had seemed natural that the two men forged a bond that went beyond being cousins. They both wanted to leave the war they had fought behind them. Now, each Maine season dictated their sport: fly-fishing, ice fishing, snowshoeing, hunting, or camping.

Jimmy settled into the passenger seat as the Hillman rolled through the small towns along Route 1A. Pete fiddled with the radio dial. The night before WPOR in Portland had reported an international AP story that caught his attention. The station had received a news wire that a British airliner was missing, a chartered twin-engine Vickers. The plane had left the airfield at Bovingdon, England, with a final destination of Nairobi, Kenya. Among the five crew members and twenty-six passengers were fourteen women and children, family members of British soldiers stationed in British Kenya, on the east coast of Africa. The plane had first landed in Nice, France, with a flight plan that would take it south over Italy. Air traffic controllers had spoken with the pilot as the plane flew over Sardinia. That was the last contact made, an hour before it was due to land in Malta for refueling. Where was the chartered Vickers?

There was no update on the plane, but the radio crackled with news about Elizabeth Taylor. The violet-eyed actress was to leave Los Angeles in the morning and fly to London where she would marry British actor Michael Wilding, her second time at the altar. She would celebrate her twentieth birthday a few days later. Waiting for the finishing touches to her gray wool wedding dress trimmed in organdy had delayed her flight for two weeks. Uninterested in Hollywood babble, Pete turned off the radio. He felt an affinity to the soldiers in Africa who were waiting for their wives and children to arrive. He himself had been a British soldier on that continent. He remembered the hunger, the swirling desert sands, and the extreme heat. The Siege of Tobruk. Too many men had died. It seemed a lifetime ago, and yet it was often present in his mind.

"There's the turnoff," Jimmy said, and pointed ahead to the sign. BRANCH LAKE. The lane called Hanson's Road ran three-fourths of a mile through a corridor of evergreens down to the northeast corner of the lake. Wind had swept away all but a few inches of previously fallen snow. The car would have no trouble driving out onto the ice, which was nearly two feet thick and safe to carry the weight. Branch Lake was known for its sportfish. Since the late 1800s, it had been replenished often with salmon. But there were a variety of fish to be taken winter and summer, including bass, pickerel, togue, perch, and smelts. All around, it was good fishing and many a trophy-winner had been pulled from those waters.

Jimmy and Pete wanted lake trout. That meant shallow water. On the previous trip they had fished for trout around the two small islands at the top of the lake and caught only a few. Now they decided on a spot they had visited before, four miles down the lake and known as the Shoals, or Narrows. The width there tapered to a thousand feet, shore to shore, a favorite hangout for trout. The sturdy Hillman drove easily across the surface, its tires crunching down four inches of top snow. Jimmy pointed to an area that looked promising, a couple hundred feet from shore.

While Laddie explored the nearby woods, the two men gathered dead boughs and twigs to kindle a fire. Then they pushed over the remains of a dried cedar tree, easily downing it. They dragged it over the ice to a spot a few yards from the car. They wouldn't run out of wood on a lake in Maine. They could keep a proper fire burning all day. When it was going nicely, they took turns chiseling through the thickness of ice to drill the fishing holes. Then they fit their hooks with live bait and lowered them into the cold water.

THE TARDIFFS ON VARNEY MILL ROAD

At six o'clock Phillip Tardiff filled the furnace in the basement with heavy chunks of hardwood. The fire burning well, he went back up to the main floor. In the hallway near the chimney he tugged the chains that would adjust the damper and allow more heat. Hazel would be up soon and then the kids. Heating a Maine house in winter, especially one built in the mid-1800s, was a steady job. And money was tight for the Tardiff family. The coal truck made deliveries when needed. But to supplement the coal, Phil hooked his tractor to a utility wagon each autumn and pulled it up the cow path and into the woods. With David helping, they gathered fallen hardwood limbs, the larger the better, and piled them into the wagon. Smaller twigs worked well for kindling. Phil had a steady paycheck from Bath Iron Works as a welder. But that salary had to support a family soon to number six. The Tardiffs had learned to cut corners. Before he met Hazel, he had been a barber in Portland. To save money, he now cut his own hair as well as David's. And Hazel's friend from nursing school gave her and the girls home perms when needed. Every saved penny helped.

Hazel found her husband in the kitchen cooking bacon and eggs. The kids had scrambled down the stairs and were jockeying for a place over the warm register.

"I'll make dinner, too," Phil offered. He was something of a disciplinarian as a father, but he was also their rock. He knew Hazel would appreciate his help. And this would be their last Sunday as a family of five.

"Spaghetti and meatballs!" said Mary Lou. The children had watched him make the meatballs the day before. The sauce would soon be simmering on the stove in a huge pot. Sunday dinner was at noon.

"And crepes tonight," Phil added.

Crepes covered in brown sugar were always expected on Sunday nights. They were his specialties, passed down from his French-Canadian mother.

"It's snowing!" David said.

"I hope there's no school tomorrow," said Mary Lou.

Hazel glanced out the kitchen window. Snowflakes were fluttering like moths over the gray waters of the Kennebec River. Canceling church had been a good decision. She would call Dr. Hamilton later to touch base. She was so near her time now. She had felt the baby move often as if impatient to join the family. Thank heavens this storm would pass over them. In the morning she would pack her suitcase and stay with her folks in town. Virginia Hamilton wouldn't be the only one relieved.

"We'll see what it amounts to," said Phil. "It'll take more than a few snowflakes for school to be canceled." He stubbed the cigarette he had smoked while cooking breakfast. With Hazel often smoking, too, the kids were accustomed to a thin gray veil hovering in rooms of the house where their parents had been.[8]

Hazel's plan was to call Bath Iron Works if the baby decided to arrive during Phil's workday. His boss would send him home to drive her to the hospital. He could then read a magazine in the waiting room with any other expectant fathers. But if the baby came at night, he would already be home to warm up their Buick. Sitting outside on the inclined driveway, the car was like a member of the family. It was a green 1946 Roadmaster with swept-back fenders and a bombsight hood ornament, a design inspired by the recent war. The chrome radio in the center of the dashboard was the size of a lunchbox. Phil drove the huge beast so slowly that the kids went crazy with impatience, often hollering from the backseat, "Go faster, Dad!" Unless the road turned icy, Hazel was confident they could make the four miles easily, even in a few inches of snow. Driving the Buick Roadmaster down Varney Mill Road was like driving a Sherman tank.

BILL DWYER, 19 BATH STREET

It was a few minutes past eight o'clock when Bill left his small house and turned left on Washington Street. He had already eaten his breakfast an hour earlier. Sundays were when he missed Nellie most. A morning walk to grab a newspaper at Hallet's Drug Store had now become a habit. Long retired, it was one way to keep his bones active.

On his right sprawled Bath Iron Works, his former workplace as a riveter. Hearing the weekend hum of pounding and grinding by the work crews brought back old memories. He and that shipyard had been born in the same year, 1884. Its founder, Thomas W. Hyde, whose family was from Bath, had become a brigadier general in the Union Army by age twenty-three. He was present at battles like the Second Bull Run, Antietam, and Gettysburg. He was with General Sedgwick at Spotsylvania when the general said, "They couldn't hit an elephant at this distance," seconds before a sharpshooter's bullet cut him down. When Lee surrendered to Grant at Appomattox Court House, Thomas Hyde was there. After the war he returned to Maine, purchased Bath Iron Foundry, and employed seven men. Interested in politics, he became a state senator and later mayor of Bath. In 1884, he reopened the foundry as the Bath Iron Works and became known nationally as a great shipbuilder. Generations of Maine workers spent their careers in the company Thomas Hyde left behind.

At the corner of Front and Centre streets he could look across to where the old custom's house once stood, where the Angel Gabriel had blown his trumpet on the wooden sidewalk a hundred years earlier. At Hilyard Drugstore, he noticed a flurry of excitement. Bath residents were boarding a bus headed down to Boston Garden for the Ice Follies matinee that afternoon. The event had been the talk of the town for some time, especially since the famous Swiss skating duo Frick and Frack, dressed in lederhosen and performing spectacular acrobatics, were on the program. With the last passenger on board, the driver closed the jackknife doors and the bus rumbled off. Other residents were waiting for that coming Saturday to see the Follies. For $23.05 each fan would receive round-trip transportation, a ticket to the evening show, and twin beds for two nights at the Bradford Hotel in the heart of Boston. It was an expensive undertaking, but the show was world famous.

Turning down Front he saw the clock at Hallet's towering above the street like a giant green tree. It was snowing lightly as he opened the door and stepped inside. As Bill paid for the Sunday paper, he read the bold headline: SNOWBOUND 100 MAKE TREK OUT! All of Maine had been relieved to hear this news the day before. They had been praying the woodsmen and their animals would come out safely before the next storm arrived. As he left the drugstore with the rolled newspaper under his arm, he met John Tourtillotte about to enter. John was a young man who had done well for himself by joining the military. After years of service, he was now a captain in the National Guard.

"How's that arm, Turtle?" Bill asked. It had been John's boyhood nickname.

"Ready for a game of baseball," John said, and smiled.

Bill buttoned his coat against the chill. Through the drugstore window he saw John placing an order at the food counter. He turned and walked back down Front Street, leaving the clock behind. When he was foreman of the volunteer fire company, they had arrived at Pleasant Street to find a dozen teenaged boys gathered excitedly around the trunk of an oak tree. Johnny was up in the tree, his arm painfully lodged in a fork where he had reached to retrieve their baseball. The firetruck ladder was extended and Bill had climbed up to help the boy free his arm. That had been two decades earlier. He missed the notion of being needed now and then. He missed the excitement, and the urgent cry of the firetruck's old bronze bell.

Before he arrived back at 19 Bath Street, Bill stopped at the corner store and bought some bread and milk. He would make himself a tuna sandwich for lunch and then let Snooky have what was left in the can. It was the cat's usual Sunday treat.

THE NEW YORK SAILOR

Paul Delaney slicked his hair back with a wet comb and put on a clean blue shirt. It was almost noon when he drove past the leafless yellow birches and the numerous shrubs covered in snow, and headed the car north. When he dead-ended into Route 1, he would turn left toward Ellsworth. The car's heater worked well. Paul tapped an unfiltered Chesterfield out of its pack, lit it, and then settled back for the drive ahead. Bar Harbor was only forty-five miles from the base and the matinee movie was not until three o'clock. But on those narrow winter roads it would take over an hour to drive safely. That it wasn't his car was another reason to be cautious. He turned on the radio and rolled the dial past a couple stations. He was in search of some good music like "Bonaparte's Retreat" by Pee Wee King, or "Harbor Lights" by Guy Lombardo. But "Mona Lisa" was still his favorite song.

It was a slow drive through rural areas like Sullivan and West Sullivan, crossing over Taunton Bay, then through the small village of Hancock. It was well-known among the crew back at the radio receiver station that Hancock Point was where two spies, one a defected American and the other a German, came ashore in a rubber raft in late November of 1944. A German U-1230 sub had brought them into Frenchman Bay. Carrying heavy suitcases and wearing coats too thin for the cold weather, they made their way to Boston and New York before being captured and sentenced to death, sentences later commuted to life imprisonment by President Truman.[9]

The incident was still fresh in military records and in the minds of local residents. But old-timers could remember back to World War I when Ernesto and Alessandro, the Fabbri brothers at Otter Cliffs in Bar Harbor, had been accused of spying for Germany. GERMAN WIRELESS AT MOUNT DESERT? the *Bangor Daily News* had asked. People had to be diligent. A wave of suspicion, the Red Scare, was now sweeping the country after Senator Joseph McCarthy's recent claim that 205 Communists were walking the hallowed halls of the country's State Department, fifty-seven of whom he could identify by name. That new neighbor asking to borrow a cup of sugar could easily be a Commie.

At the town of Ellsworth, Paul took Route 3 south. It would carry him onto Mount Desert Island and into Bar Harbor. The closer he got, the more anxious he became. When he first met Mona three weeks earlier and asked if he could take her to a movie, she seemed pleased. Bar Harbor was famous in the area for its art deco theater called the Criterion, on Cottage Street. Paul and Jeffrey had driven over there a few times already to see a movie and check out the pretty girls. Mona had been sitting with her friends in Harris's Soda Shop, the local hangout. It's where all the kids went for ice cream sodas and burgers after a film.

Paul wasn't excited about the movie that was on the matinee bill. *Golden Girl* starred Mitzi Gaynor. But Mona was thrilled. "It's about the life of Lotta Crabtree," she had told Paul, her eyes bright. "She was only sixteen years old and the Civil War was going on and yet she performed all over the country and became a big star." He would have preferred a guy's film. There was one coming that weekend with Jimmy Stewart in the leading role. But sitting next to Mona for any movie and maybe holding her hand was enough for Paul. He rolled the radio dial again. Hearing Nat King Cole sing "Mona Lisa" would be hitting the bull's-eye. But all he found was static. He turned the radio off and concentrated on the narrow road.

By the time he saw the Bar Harbor welcome sign there was no sun visible and the sky had clouded over. Scattered snowflakes were melting on the windshield. That was fine so long as it didn't turn into a storm. Otherwise, it would depend on Mona how late he left that night to return to Winter Harbor.

LEAVING MONHEGAN ISLAND

It was not yet noon. Harland and Jimmy had the thirty-foot *Sea Breeze* loaded. The fifty lobster crates were stacked on top of each other on the stern deck, an area eight feet wide and twelve feet long. Five thousand pounds was a heavy load for a boat that size. Harland had paid his fishermen clients with checks. He and Jimmy Haigh would settle their own bill back on the wharf at Port

Clyde. There was no time to eat a quick lunch after all. The sky was now dark and foreboding. Down to a man, the Monhegan fishermen had looked out at the sea and said, "Don't risk it, Harland." Already the winds were bearing down and the ocean swells growing larger. Since those winds were coming from the northeast, Harland's plan was to get in the lee of Burnt Island on its western side. They would be better sheltered there. He would then cross to Marshall Point. Following this route would take them past the Old Man Ledge whistling buoy.

"Stay here until the storm passes," the Monhegan fishermen, like a Greek chorus, advised. It was the kind of advice you gave a friend but didn't always follow yourself. All seasoned fishermen took a chance on the ocean now and then.

There were considerations all around and Harland and Jimmy discussed them. If the storm didn't pass quickly it would mean spending the night on Monhegan. Jimmy needed to get those lobsters back to Portsmouth for his Monday morning clients. He had a family to support and a business to keep afloat. Bills had piled up. But so had Harland's. And now that Alice was pregnant, they would soon be a family of four. Maybe Haigh pushed to get back, having the most to lose. In the end, the decision belonged to Harland Davis alone. He owned the *Sea Breeze*. He was no fool. Jimmy Haigh was no fool. They both believed they could beat the northeaster to the mainland. They put on their life jackets and said their goodbyes. The fishermen stood on the wharf and waited until the *Sea Breeze* disappeared from view.

What surprised the two men was how well-formed the storm was, and how fast it found them. It seemed to arrive out of nowhere. They were almost halfway to Port Clyde when they met its true wrath, the sea beating against the boat. It was now half past noon. The horizon had disappeared in a cloud of wet snow that was closing in on them. The running lights were useless to cut through the spray. The boat struggled against the gusts of wind as they grew stronger, slamming its starboard side. Waves washed over them, each time filling the boat with more seawater. The only protection Harland and Jimmy had was inside the tiny cab. At the helm, Harland gripped the wheel.

He needed to maintain enough headway to keep the bow up. As the *Sea Breeze* took on more water, there was little separation between gray ocean and grayer sky. It was now impossible to see beyond a few yards and with each blast of snow-filled wind that distance was reduced to a few feet. Each time a wave crashed over the bow, Harland fought to stay on course. He needed to get in the lee of Burnt Island, but he was being blown off course by winds of gale proportions. They rocked the boat, hitting it broadside. If it flipped, they were goners.

BRANCH LAKE

When it was time for lunch, Jimmy and Pete sat in the car, one in the front seat, one in the back, both doors open and their feet on the snow. It was not only comfortable, they were sheltered from any wind blowing across the lake. The sky had clouded over. A few flakes were falling, but nothing worrisome for Maine. Maybe southern towns on the coast were seeing some of the incoming northeaster as it passed by, but so far not Branch Lake.

Laddie lay like a sphinx on the ice, watching the edge of the woods for signs of a wandering rabbit. The fishing had been decent so far, a few perch, a few smelts, a couple of nice-sized lake trout. On previous trips, they even cleaned a few on the spot and fried them in the frying pan Jimmy kept well greased. Pete had steeped his necessary cup of tea by heating water in an empty can attached to a wire, then tossing in some tea leaves. He could have brought a thermos of hot water and teabags, but there was something more satisfying in using the fire. When they finished eating, he packed tobacco into his pipe, an extra mild Cavendish, to enjoy an after-meal smoke. The gray spiral from his pipe drifted with the wind out over the lake. Their lines were still in the water, the tip-ups moving slightly as the live bait swam below the ice, circling.

When it was time for the twelve o'clock news, Pete turned on the radio. The 1952 Olympic games were half over in Norway, but controversy was raging over Soviet athletes competing if they had been German sympathizers. One of Norway's top speed skaters was disqualified for having collaborated with the Nazis. And should Germany be allowed in the games at all? Norway hadn't forgotten the German occupation of that country. This sort of news troubled Jimmy Morrill and Peter Godley. It was as if the brutal war they had fought on the battlefield was now being waged in sports and politics.

Then came an Associated Press breaking report. A search plane hunting for the Vickers had left Malta and flew north over Sicily, encountering thunderclouds and strong winds near the mountainous terrain. Banking along the rugged slopes of Monte delle Rose, the pilots spotted the downed plane near the 4,600-foot peak. Before turning back due to bad weather they glimpsed a body lying next to the wreckage. They also saw survivors waving frantically. At least this was what air controllers on Sicily's opposite coast reported hearing in an intercepted message from the search plane. Ground rescue parties from nearby Sicilian villages were already on their way to the top of the mountain in search of survivors.

"Those poor fellows," said Pete. Jimmy nodded. The British soldiers waiting for their wives and children to arrive in Nairobi wouldn't have an answer for hours or maybe days as to who was alive and who wasn't.

"We got another one," said Jimmy, and pointed at the tip-up that had suddenly shot downward, pointing at the drilled hole.

OLD MAN LEDGE

No hawk hangs over in this air:
The urgent snow is everywhere.
The wing adroiter than a sail
Must lean away from such a gale . . .

—"The Snow Storm" by Edna St. Vincent Millay

The ocean was threatening to bury them. Water was rising in the narrow spaces around the lobster crates and streaming through their slats. Several stacked on the top tier had already blown into the sea. There was a pail Harland kept on the boat for bailing, but the act would be useless. Even if they could get to the bilge, it would be like using a teacup. As gray seawater whipped over the side of the craft, they knew they had to lighten the load. While Harland stayed at the helm Jimmy began lifting crates and throwing them overboard. To hell with money. This was now about living or dying. Wind and waves slammed the crates back against the sides of the boat.

There is no doubt that the time came quickly when Harland Davis and Jimmy Haigh realized the odds were against them that day. They knew they were a couple miles from any land. They knew the boat was overloaded with lobsters. They knew the temperature of the water was cold enough to kill them quickly. Even wearing the life jackets, it was a matter of minutes before hypothermia would set in. But a life preserver increased their odds of being rescued. They were now standing in water. It spilled over the tops of their boots and their feet and toes grew numb. Icy wind lashed at their faces and stung their eyes. All around them, the granite-gray sky held nothing but water and snow.

No matter how many crates Jimmy threw overboard the boat was submerging, unable to carry both the heavy load and the wind-driven sea. The frigid water crept higher. A plunging wave dipped the bow, shorting the electrical system. The running lights blinked out. Their gloves soaked and their hands freezing, the boat was sinking beneath them. A wave slammed a crate back against Jimmy's right hand, breaking his wrist. His eyeglasses were knocked off and swept into the sea. With the stern of the boat now mostly underwater, he and Harland clung to the sprayhood. With little sensation left in their hands the effort was useless. Maybe in those seconds before the mind-numbing cold set in, Harland shouted, "Hang on! They'll find us!" Maybe Jimmy shouted it. Then a new wave carried them both away from the boat and into the sea.

Human beings are not designed to withstand freezing temperatures. Their bodies reacted immediately, a *cold shock response*, by hyperventilating. The life preservers held them aloft as the churning waves pulled them farther apart. In a few minutes their thought processes would become jumbled and confused. But while they hoped, while they prayed, they floated like human corks in an uncaring sea.

BURNT ISLAND LIFEBOAT STATION

I know what you're thinking. The three of us will
probably die trying to save one guy who will die
also. Get in the boat—we have a job to do.

—Warrant Boatswain Alston Wilson,
Great New England Hurricane, 1938.

The coast guard station on Burnt Island sat five miles from Port Clyde and seven miles from Monhegan Island. Under the command of Weston

Gamage Jr., a boatswain's mate first class, the day began with the usual Sunday drill. At 0800 hours it was morning colors, with the hoisting of the flag as Weston and his four crewmates saluted the Stars and Stripes. This was followed by a routine inspection of the station and grounds. The radio and other equipment were in working order. The winch engines on the boats were checked and were in good running condition. With these duties finished, they were free to observe the Sabbath, with each crewman doing a four-hour shift in the lookout tower on the southern tip of the island. By noon it was snowing with strong winds lashing the water. Visibility was less than a mile. This was not a day for small craft to be on the ocean.

At 1:15 P.M. the telephone rang at the station. It was Riley Davis of Pleasant Point calling about his son, Harland. He had left Monhegan Island just before noon with a load of lobsters. With him was a buyer from New Hampshire. They were in a thirty-foot boat with the registration number 1-L-668. Could the coast guard be on the lookout for Harland's boat? Weston assured him he would. Through the windows of the station, visibility was now less than a half mile. Even an empty boat would be in danger in those swelling seas. Gamage figured that if Davis was coming from Monhegan he may have planned to pass Georges Islands on the western side in order to take advantage of the sheltered lee. If he got in trouble near the Old Man buoy whistle, the northeast winds would push the boat west of it. That was the best place to start looking.

There were two boats assigned to the Burnt Island station. In this weather, they would need the thirty-six-foot self-righting lifeboat with CG-36506 painted on its side. Should it turn over in the worst of storms, it would immediately roll back to an upright position.

"Get out the boat," Weston said to his crew.

It would take them fifteen minutes to mobilize. Old Man Ledge, lying in shoal waters, could present a danger even on a good day. And now the sea was reaching gale winds of forty-five miles per hour.

At thirty-four years old, Weston had a wife and young son at home. But it was not the time to think of his own safety or even that of his crew. That's not how the coast guard operates. Their formal motto might be Semper Paratus (Always Prepared), but there was an informal one they all knew well. "You have to go out, you don't have to come back." That message was in Gamage's blood. He had been born and raised on Maine's ocean, in Southport, as had his father who also served in the coast guard.

With Weston Gamage at the wheel, the craft sped away from the Burnt Island station and into the storm. It was headed toward the buoy at Old Man Ledge, one and a half miles away. Gamage never knew what to expect until he arrived on the scene. Would they come upon Harland's boat making slow progress given the storm? If so, they would stay as close as they could in those seas and escort it safely to Port Clyde. If the seas worsened, they would guide it instead to the Burnt Island station. The two men could sit out the storm there, with cups of hot coffee and donuts. Or if the boat had stalled, Weston's crew would attach a tow line. Were Davis and his companion injured in any way? Was the boat taking on water? Had it capsized? Were they wearing lifejackets?

The questions were many. The answers lay somewhere in the ocean near Old Man Ledge. One half mile west of the buoy, Gamage and his team came upon the sprayhood of a boat that was mostly submerged. Beneath the water, visible in the spotlight on the starboard side, was the number 1-L-688. Wooden crates filled with lobsters bobbed in the water around the boat. There was no sign of the two men. Weston began a search of the area, circling as the spotlight combed the waters, widening his path with each pass. Five minutes later, he spotted a man floating in a life jacket.

FISHERMAN'S PRAYER

O, God, thy sea is so great and my boat so small.

Hypothermia occurs when the core body temperature of approximately 98.6°F drops below 95°F. The heart, the nervous system, and other organs notice this change and begin to work abnormally. The first sign of hypothermia is usually uncontrollable shivering as the body's automatic defense system kicks in and struggles to warm itself. Coordination and strength are lost as blood moves away from the extremities toward the center of the body. As the rate of metabolic processes slow down, there will likely be slurred speech and mumbled words, as if in a state of drunkenness. Breathing is shallow with only some of the breathing muscles being used. The pulse weakens. Drowsiness can follow and coordination become clumsy. The mind fights to make sense of the situation, but there is confusion and memory loss.

After the loss of consciousness, death may occur quickly. Each person is different. A lean body has less insulation to protect it than a stout body with its stored fat. An attempt to swim will use up the last supply of energy and may shorten the survival time. In water that is approximately 36°F, as it was on February 17, a human being can survive ten or possibly fifteen minutes before the muscles weaken. When waves knocked Harland Davis and James Haigh away from the *Sea Breeze*, they may have tried to swim back if their limbs were still usable by then. At that moment in time, the exposed sprayhood was the only solid thing they possessed, their only refuge in an furious sea.

Because dying is a unique experience to every human being, each one confronts it differently. Those who do not survive have no way of sharing that final experience. But survivors in near-death situations have reported that the human brain becomes distorted with random thoughts, even a sense of self-atonement. "How could I have let this happen? What will my friends say?" Sometimes there is a sense of curiosity. "So *this* is what it feels like to die?" Illogical, even routine thinking might occur. "I can't miss my doctor's appointment next week." Or, "I won't know if my team wins the World Series." For still others, they relive a speeded-up film clip of their lives.

Perhaps Harland Creamer Davis thought of hauling lobster traps by hand at his father's side, a boy learning the trade. An autumn day, the air sunny and crisp. The morning ocean shimmering with diamonds. The poplar near his house, its leaves curling in the wind like silver dimes. The *Sea Breeze*. The sound of hammers, the smell of freshly sawed lumber. His wharf standing proud, a red handkerchief waving in the breeze from the derrick. The smell of Alice's perfume. Little Carolyn, running into his arms when he knelt for her. The baby Alice would have that summer, a child he would never hold. He had wished for a son. He would want his family safe. He hoped someone would take care of them if he didn't come home that day.

James B. Haigh had lost his older brother three years earlier when a blood clot traveled to his lungs. His mother had buried her oldest son, who was only forty years old. Now her second son would be lost. Ellie. They had faced so much together, he and his wife. The baby who never had a chance, stillborn and deformed at nine months, part of its brain missing. The doctor wouldn't let Ellie see her boy. Out of pain, renewal. The company was growing. Hard work. Patience. It had all been in their grasp. What would they do without him, his wife and daughter alone? Barbara Ann was Daddy's girl. Any man wants to walk in his front door again and see the faces of his loved ones. He would do what always made them smile. He would kiss Ellie first, and together they would scoop Barbie up into their arms for a family hug.

The sea was a mass of cold waves beating at their bodies. There was no heaven overhead, just gray sky pressing down, a mountain of snow and wind and water. Whatever their last thoughts were, both Harland and Jimmy would have heard, at least for a time, the sound of the whistling buoy at Old Man Ledge, its constant cry produced by the rise and fall of the sea. Hypnotizing. The voice of a siren luring sailors to those rocks. Over and over again. The whistling buoy, like a mother's lullaby.

RIDING ON HOPE

Holding the boat hook—a six-foot pole with a nine-inch rounded brass hook at its end—a member of Weston's crew leaned over the side of the lifeboat. After a few attempts, he was able to latch the hook onto the life jacket of the man in the water. The boat pitched beneath the crew, the waves persistent. Quickly, he pulled the body up to the side where two other crew members waited. One grabbed the arms and the other grabbed the legs. They lifted the unmoving man from the freezing water and onto the deck. The man—Davis or Haigh, they had no idea—showed no visible signs of life. But that meant nothing. The practice of the coast guard under those hypothermic conditions is to attempt resuscitation nonetheless. The boat had no heater, as was typical of coast guard lifeboats. The only heated area was the engine room, but it was too cramped and dangerous to attempt resuscitation there. They turned him face down on the deck.

As Weston began an ever-widening search in the rough seas for the second man, his crew members removed the life preserver of the one before them. The coast guard's approved method of resuscitation was the back-pressure arm-lift technique, implemented just two months earlier, in December of 1951. The victim's elbows were bent, his folded hands placed beneath his face, his cheek resting on his hands. One crew member knelt at the head and placed his hands on the man's mid-back. He then rocked forward, using his weight to apply pressure on the back. This motion allowed his hands to slide up the victim's arms to above the elbows. Leaning back farther his grasp would raise the arms and lift the chest from the floor to expand it. Or at least that was how it worked. Effective or not, this manual method required a steady rhythm in application if it were to be successful.

The lifeboat swayed atop the waves, pitching the crew to and fro. Even if there had been signs of life in the man who lay on the deck, resuscitation was impossible. It demanded a steady rhythm that could not be found in forty-five-mile-per-hour gale winds. There was nothing left to do but wrap the cold body in blankets and go back to the rail, hoping Weston could find the second man. Ten minutes later, the spotlight's beam lit up another body bobbing inside a life jacket. They had found him in that churning sea. Again the hook with its rounded tip so as not to puncture the body pulled the limp man over to the boat. Beneath his life jacket he was wearing a burgundy-and-gray barn coat. Harland Davis.

It was 2:55 P.M. when Weston Gamage turned the boat toward Port Clyde. He radioed ahead that the coast guard was bringing in two unconscious men,

fishermen who had been to Monhegan. Their boat was submerged to the sprayhood. Gamage requested that firemen from Rockland be waiting with their
rescue-breathing apparatus. He also asked for a doctor and an ambulance.
As the lifeboat struggled over the waves toward shore, crew workers again
attempted to resuscitate the first man they'd brought on deck. The method
was still unworkable on those jagged seas.

THE COUNTRY DOCTOR

Born in the spring of 1880, Charles David North was a local legend. He had
graduated from the Medical School of Maine in 1907. It was the same institute
where Theodore Jewett, father to the Maine author, had studied and was also
a professor. Sarah Orne Jewett fictionalized her father in *A Country Doctor*,
published in 1884. Future generations of medical school graduates admired
Dr. Jewett. Charles North was *that* kind of doctor. When he first began his
practice, he made house calls in a horse and buggy. It was common knowledge
in Rockland, where he practiced for forty years, that Dr. North would have to
be bedbound to say no to a patient in need. And yet, in almost a half-century
of service, he rarely allowed himself to be ill. He had even worked courageously
around the clock to save patients during the 1918 Spanish Influenza.

For many years North maintained an office at his Rockland home, even
after he was too ill to travel in his midseventies. Patients walked down a long
hallway leading to the office door. They were accustomed to the sight of a cord
from his shirt pocket leading up to a plastic earplug. His hearing was long
past what it used to be, and the Zenith device went everywhere the doctor
went. Many of these patients couldn't pay for services. At the end of a day the
hallway table was often filled with fresh garden vegetables, a bag of apples from
a farmer's orchard, or wild strawberries from a nearby field. In the winters,
there were jars of canned fruits and vegetables, or homemade pies. A plucked
chicken might be left at the back door. He never turned away a patient in
need, even after he could no longer hurry to their bedsides.

Despite his admirable traits, Charles North was quite hopeless driving an automobile. He often mourned the loss of his beloved horse and buggy. As if to make up for it, he insisted on buying huge Chrysler DeSotos, always sky-blue in color. The fenders and bumpers of each were laced with telling dents. He regularly parked his car closer to the middle of the street than to the curb. Locals were amused and policemen looked the other way. After all, the imposing blue auto was immediately recognizable as Dr. North's car. And it was parked where it was because the doctor had been summoned to a patient in need.

In 1935, Charles North was made medical examiner for Knox County. He had seen more suicides and accidental deaths than he cared to remember. For a man whose passion had been saving lives, this was a job he did for his community. Just as the northeaster was hitting coastal Maine with a barrage of wet snow and gale winds, the wall phone rang at the North residence. Most patients knew by heart the doctor's three-digit phone number: 712. His grandkids always giggled when they saw him hold the phone close to the receiver box in his shirt pocket so that he could hear the caller. It was 3:00 P.M. and the incoming call was from the coast guard station. A lifeboat was on its way to Port Clyde with two fishermen aboard who were presumed dead. Nonetheless, if there was the slightest chance of revival, the code of the coast guard and the medical profession was to do everything in their power to save a life. If a miracle was possible that February day on the dock at Port Clyde, now was the time for it to happen.

THE SAILOR IN BAR HARBOR

Mona was wearing a blue sweater and a tweed skirt that reached the top of her tall boots. She had waited for Paul as planned at Harris's Soda Shop where they would eat after the movie. She smiled when he stepped inside and closed the door. He had made good time driving over. The snow was not letting up, but that was still not a problem. Unless the storm grew in intensity, he'd be fine getting back to the station so long as the plows did their jobs. The Criterion Theater was just a few yards up the street, but Mona wrapped a white scarf around her neck and put on her matching white gloves. As they walked, she told Paul again how excited she was to see Mitzi Gaynor acting out the life of Lotta Crabtree. She had even done a paper on Miss Crabtree for school, so impressed was she with such an adventuresome life. Her brown hair falling around the collar of the blue coat she wore made her look like a movie star herself.

Bar Harbor's original theater had been the Star, once frequented by Ernesto and Edith Fabbri, the Rockefellers, Vanderbilts, and other wealthy rusticators at the turn of the century. The Criterion now stood just across the street. But while the Star had been comfortably rustic in appearance to satisfy the pastoral notions of the wealthy, the Criterion was anything but. With its art deco front and inside floating balcony, it appeared out of place to Paul. He expected to see it somewhere in New York rather than Maine. Mona waited as Paul paid the twenty-five cents per ticket. Inside the lobby they stopped at the candy counter for popcorn and soft drinks. They found seats in the floating balcony and he helped her off with her coat. She was wearing just a hint of perfume. By the time the movie began and Mitzi Gaynor was singing, dancing, and flashing her petticoats at the Crabtree Boarding House, Paul Delaney didn't care if he ever saw Jimmy Stewart in another film.

ON THE LAKE

By afternoon the gray sky over the lake was fading, turning into dusk. The cousins had kept fishing as white flakes piled up lightly on the older snow. Even if it amounted to a few new inches it would not be a problem for the big Hillman. They could still drive off the lake and had done it before. If a squall moved in, it would pass over them in a couple minutes. The town plows in Maine were industrious and reliable. Jimmy and Pete would have no trouble getting back to Brewer on Route 1A. They were certain of it. Just a little longer on the lake couldn't hurt. They had another busy week ahead, Jimmy bartending at the tavern and Pete working at the shoe factory. Seeing activity behind a fallen fir tree on the shore, Laddie went bounding off the lake and into the woods. Jimmy rebaited the hooks as Pete lit his pipe.

ALICE FRENCH DAVIS

She had put off getting her hair done for days. There was always something else to tend to, either Harland needed an errand run or Carolyn help with her schoolwork. She had been experiencing morning sickness as well. Now, with Harland gone to Monhegan Island until that afternoon, Alice made plans. Her friend Martha, whose husband was also a fisherman, ran a tiny beauty salon out of her house. It was a one-chair business that stayed busy. "Sunday is fine," Martha told her. "While I do your hair we can visit." Alice could have done her own hair. She had an ample supply of metallic curlers that rendered the tight curls she liked. But getting out for some fresh air, even the short walk to Martha's house, seemed like a better idea.

It had started snowing just before noon at Pleasant Point, soft flakes at first and then more laced with wind. A winter storm in Downeast Maine was just that, unless it turned into a blizzard. Having Martha chat away as she liked to do with her customers would take Alice's mind off the weather. In a stock pot on the stove, she left a fish chowder for supper. That morning she had baked cream puffs using her own recipe for pastry cream and sprinkling them with powdered sugar. Harland and Carolyn were always asking her to make some. She wrapped up two and put them in a paper bag to go with Martha's coffee. Before she left, she checked on Carolyn, who was in her bedroom doing homework.

It was two o'clock when Alice put on her coat and boots and pulled a scarf over her hair. As she walked to Martha's house she was thinking that the *Sea Breeze* would soon be docking at Port Clyde, if not earlier. Harland would help James Haigh load the lobsters into his truck for the trip back to

Portsmouth. And then he'd be home. They would enjoy a family supper and a quiet Sunday evening.

"I got the coffee on," Martha said when she opened the door.

"I hope the smell doesn't make me nauseous," Alice said. She left her coat on a chair by the door and kicked off her boots.

"Think that storm will hit us?" Martha asked.

"I hope not," said Alice. "Harland's gone to Monhegan with a buyer. But they should be back to the mainland by now."

It was not yet four o'clock. Martha had pulled the curlers from Alice's hair and was styling it. They had already enjoyed the cream puffs as they gossiped about Liz Taylor's upcoming marriage. The sounds and smell of fresh coffee percolating drifted in from the kitchen. They would share a second cup before Alice headed home. Chatting as they were, they hadn't noticed how the storm had worsened until the door opened and Jack came in, followed by a gust of wind.

"It's really coming down," he said, brushing snow from his hat onto the mat at the door.

"That's a good husband," said Martha, winking at Alice. "Now wipe your boots."

Jack stopped in the kitchen to pour a cup of coffee.

"A call just come to the general store," he said. "The coast guard is bringing two fishermen in to Port Clyde. Their boat capsized on the way back from Monhegan."

Neither woman spoke. Alice was wondering if she had really heard those words.

"They radioed ahead for an ambulance," Jack added. "And for the firemen to bring that breathing thing they have. But a man couldn't last long in that water."

"Jack, no," Martha said quickly.

Alice didn't speak. Her brain was trying to untangle the meaning. *Monhegan. Two fishermen. Ambulance.* It was a Sunday. Who else could it be?

Noticing now the warning look Martha was giving him, Jack stood with his cup frozen in midair.

"I have to go," Alice said. Martha tried to stop her.

"Stay here," she said. "Jack will go down to Riley's house and find out."

Riley Davis. Harland's father. How would he hold up if anything had happened to his son? Alice reached for her coat.

"I need to be home," she said. "For when someone calls."

Martha tried again. "They wouldn't have asked for an ambulance, Alice, if it was too late."

But Alice Davis wasn't listening. She put the scarf over her perfectly done hair and walked home in the wind and snow to wait for a phone call or a knock on her door.

THE PORT CLYDE WHARF

The firetruck from Rockland had pulled up and parked next to the ambulance, a "combination car" that doubled as a hearse for Thomaston's Funeral Home. Dr. North's sky-blue DeSoto was last to arrive, the doctor bundled in his usual black winter coat, dark gray fedora, and a knitted woolen scarf about his neck. The twenty-one miles down from Rockland was a challenge, even for the DeSoto. Snow was now filling the narrow roadways and a heavy wind blowing it back into place once the plow had passed. This was something the lumberjacks up in Aroostook County had learned well the week before. Charles North feared he would be busy if this kind of snow kept falling. It was four o'clock with darkness coming on when the running lights on the CG-36506 emerged from the grayness of sea and sky. It had taken Weston Gamage an hour to buck the swells over eight and a half miles of ocean.

When the firemen jumped down from the truck to get the breathing apparatus set up on the pier, Dr. North opened the door of his car and got out, his medical bag in hand. He reached into his pocket for his listening device and turned it on. The batteries required to operate the miniature vacuum tubes were expensive and wore down quickly. Plus, the doctor hated the sound of his own shirt rustling in his ear each time he moved. It had been forty-five years since he graduated from medical school, and yet every time he signed a death certificate for a man or woman he thought of the larger picture, of the loved ones who would mourn the loss. The brief call from the coast guard station had indicated that the victims were young fishermen.

North watched as two firemen carried the breathing contraption from the firetruck, each lifting a side. The ambulance driver then pulled his car up next to where the firemen were kneeling to provide a shelter from the wind. A fireman unlatched and opened the case. These existing Emerson resuscitation units came with heavy tanks of oxygen that powered a device to force air into the lungs. They were so bulky that two strong men were needed to transport one from an ambulance or firetruck to the victim. While they were better at times than a patient receiving no oxygen, especially smoke-inhalation or drowning victims, they came with built-in flaws. The major one was that the unit depended on a patient's ability to physically breathe in the provided air. Anyone in dire need of *rescue breathing* benefited little to nothing.

It was high tide, so Gamage had no difficulty easing the boat close to the pier. His crewman then tossed the mooring line to one of the firemen. Several fishermen who lived in nearby houses had seen the ambulance and firetruck and walked down to the dock. They knew Harland Davis had left Port Clyde that morning and hadn't returned. They knew him as a friend and a well-respected fisherman. No one said aloud what they were thinking. *If this can happen to Harland, it can happen to anyone.*

Harland and Jimmy had been strapped to litters before the boat reached the wharf and now the first was lifted up to the hands of the firemen reaching down. As the second litter was being hoisted, Dr. North knelt next to the first body. Years of experience told him it was useless to employ the breathing apparatus. Still, he reached a hand inside the frozen life jacket and slipped it beneath the collar of Jimmy Haigh's coat. The body was rigid. He pressed a finger into the groove near the windpipe. There was no pulse.

"He's gone," the doctor said. A glance at Harland Davis's body was enough, but Charles North undid the top button on the coat and repeated the action, a finger placed near Harland's windpipe. He shook his head. "They were probably gone before they were pulled from the water." He would do a more thorough examination at the funeral home.

"That's Harland Davis from Pleasant Point," one of the Port Clyde fishermen leaned in to say, his voice breaking.

Dr. North searched Jimmy Haigh's pockets for a wallet and pulled one out. He rifled inside for a driver's license.

"James B. Haigh," the doctor read aloud. "Portsmouth, New Hampshire." They had all noticed the red Ford truck parked on the wharf with New Hampshire license plates.

A police car rolled down to the wharf, its blue light flashing. It was Willard Pease, the Knox County sheriff, who had been out on a call when Gamage radioed for assistance.

"I'll need to do up my reports," North told the sheriff. "But it looks like exposure, not drowning." Since it was obvious there was no foul play involved, North would rely on the history provided him by the coast guard. Both men had been found above water in lifejackets, and in hypothermic conditions.

Willard Pease nodded, his face grim. "We'll need to notify the families," he said.

Snow and wind were now sheeting across the wharf. The driver from Thomaston Funeral Home opened the back door of the ambulance, which now became a hearse. The firemen helped load the two litters into the back, fitting them side by side. Dr. North would follow the hearse. Having red taillights as beacons on the road ahead of him was a bonus. The firetruck pulled into line behind the hearse and the big blue DeSoto. Harland Creamer Davis and James Belmont Haigh began the trip north to Thomaston, where Haigh had been drinking hot chocolate behind the wheel of his truck early that morning before he turned south to Port Clyde. And where Harland had graduated a decade earlier from high school as class valedictorian.

THE WAIT

Alice and Carolyn sat on the sofa, Alice brushing her daughter's long hair. She could almost be her little sister except Alice loved being a mother. Alice French Davis had married for the first time when she was thirteen and a half years old and over three months pregnant. Her nineteen-year-old groom had soon gone off to the military. When her daughter was born Alice named her Carolyn, after her favorite doll. It was a marriage and a place in time she did not want to think back on, or discuss with family and friends. It was another part of her life, and it was gone. Her life was now with Harland Davis, handsome, kind, proud of his vocation and his new family. Carolyn loved him as much as a child could love a father.

Alice remembered the day the two girls had moved into Harland's tiny trailer. "We can save money this way," he told them. "My business is doing good. Before you know it, we'll build a nice house on a big piece of land." Then he had proudly shown Carolyn her own bedroom, so small only a cot would fit. He let Alice furnish the place so she would feel at home. She picked out a sofa in Thomaston, dusty blue with a matching armchair. And a yellow-and-gray Formica-topped kitchen table with vinyl-and-chrome chairs. There wasn't much room for any more furniture.

Finally, with precious minutes passing, Alice knew it was time to tell Carolyn. She put the hairbrush down on the arm of the sofa.

"I think something happened to Harland," Alice said.

"Is he all right?" Carolyn asked. She was not yet eight years old and was already used to an adult world filled with readjustments. But she had settled perfectly into Harland's life. He was the only father she knew.

"Sweetheart, I don't know," Alice said. "But I don't think so."

They had asked for an ambulance and the fire department to come with their resuscitator? She replayed in her mind the coast guard message as Jack had relayed it. And then Jack's own declaration. *But a man couldn't last long in that water.*

When Carolyn's eyes teared up, Alice brought her daughter into her arms. They held each other, waiting. The knock on the trailer door was a sound neither would forget.

"Mama?" Carolyn said, looking up to her mother for guidance.

"Go wait in your bedroom, honey," Alice said. "I'll be in soon."

There were times she had pleaded with him. *Give up fishing, Harland. But, honey, it's my job. Harland, there are other jobs. None that are in my blood, Alice.* And so she would support him because she loved him and she understood. And later, when word came of another fisherman drowned somewhere off the Maine coast, it would start all over again. *There are other jobs, Harland. None that I love, Alice.*

Alice went to the mirror in the living room. She had always wanted to look her best for Harland. He liked her hair curled. In truth, Harland liked what Alice liked. When she got home from Martha's she had put on the small pearl necklace she wore on their wedding day. And changed into her silk blouse. It was a light pink that buttoned in the back, with pin tucks and embroidered roses on the front. She had admired it at Burdell's Dress Shop when they were Christmas shopping, a hint for any husband wondering what to get his wife. When it was not in either of the two gifts she opened on Christmas morning, Alice said nothing. It was expensive and money didn't grow on trees. On their first anniversary three days later, he came into the bedroom carrying a tray with toast and coffee. He knelt and pulled a box from under the bed, gift-wrapped from Burdell's. That had been just six weeks earlier. Now it seemed ancient history.

As she expected, it was Riley Davis, her father-in-law, descended from a long line of Davis fishermen. There was nothing he needed to say. Words were unimportant. The pain that lined his face told the burden of his grief. Harland had always brought his father joy, learning the trade at his side, learning nets, and traps, and boats, a boy in love with the ocean. The truth as Riley knew it, and as Alice grew to learn, is that there *are* no other jobs when fishing is in your blood.

THE *SEA BREEZE*

It may not have been the best decision given the circumstances of the sea that day, and with the northeaster now pounding the coast. But once Dr. North had declared the two men dead, Weston Gamage and his crew fought their way back to the area west of Old Man Ledge where they had been pulled from the water. Weston hoped to spot the sprayhood. If there was a chance of rescuing the man's boat, it would be one less burden on Harland's family. Fishermen were not known for ample savings accounts or emergency funds. They were self-employed seasonal workers whose foremost and immediate thoughts were of supporting their families. The floodlight searched the darkness of the roiling seas. But the *Sea Breeze* had gone under.

It's possible that Harland Davis named the *Sea Breeze* in memory of his days at Thomaston High School. While it was not the most singular name for a vessel on the ocean, it *was* the name of his school's yearbook. He had excelled in school. It was in cement, between the covers of the *Sea Breeze*. The 1941 graduating class had shared their future dreams on one page, and left their youthful autographs. Harland Davis, whose favorite saying was "I'll tell ya!" had no trouble listing his chosen profession: *Lobster Fisherman*.

With visibility now at zero and the gales more treacherous, Weston Gamage and his crew made their way back over the waves to the Burnt Island Station. By six o'clock they had logged in and changed to dry clothing. Little was said as they ate a warm supper. It's not a good day when the coast guard fails in its duty to rescue men in peril from the sea.

PORTSMOUTH, NEW HAMPSHIRE

Ellie Haigh had been watching the street. Wind and snow spiraled around the streetlight on the corner. New England was being hit hard with a northeaster and that was not what her husband expected when he drove to Maine early that morning. Where was Jimmy? It was six o'clock. He was due back two hours ago. She didn't know who in Maine to call. The phone number Jimmy had for Harland Davis he had taken in the truck with him in case he needed it. There could be many reasons for the lateness. A flat tire. The truck breaking down. Maybe he hit the storm head-on and had no choice but to sit it out at a gas station. But the telephone would have rung at 55 Gates Street, a reverse-charge call to let Ellie and Barbara know his situation. Each time it did ring, the sound brought both expectation and dread. But it was usually Jimmy's worried mother in nearby Dover. Her first son had died just three years earlier and she had become overprotective of her remaining son. Or it was Earle Sanders, who worked for the company. *Is Jimmy back yet?*

Barbara was doing homework on the desk her father made in high school, her dog asleep on the floor nearby. They had eaten supper without him and she had helped her mother with the dishes. A covered plate of food was set on the stovetop to keep warm.

"We'll hear the truck any minute now," Ellie said. She didn't want to alarm the little girl. "Get ready for a three-way hug."

"Maybe the police stopped him for a Moxie," Barbie said. It made Ellie smile, a good way to break the tension.

"In this snowstorm?" she asked. Jimmy Haigh was not just well-liked by the local fishermen he dealt with—one insisted on calling him "Captain" despite Jimmy's protests—the local and state policemen knew him well. Now and then on a hot summer's day a state trooper would pull the red Ford truck over in case Jimmy Haigh was driving instead of Earle Sanders. That meant there would be an extra cold Moxie in a bucket of ice on the seat beside him.

Ellie had gone back to the kitchen when someone knocked on the front door. Everyone who knew the Haighs well came to their side door. Barbie ran to the bay window and peered out. In the snow falling beneath the porch light, a police officer stood on the steps. She undid the lock as Ellie waited paralyzed in the kitchen doorway. The policeman stepped inside, brushing snow from his shoulders.

"Tommy, what in the world?" Ellie asked. "Is something wrong?"

Tom Barlow had been her classmate in high school. She saw him now and then at community functions she attended with Jimmy. Before he could answer, Ellie realized why her school friend, *a policeman*, was standing in her house in the middle of a snowstorm.

"It's Jimmy!" she cried. "He's hurt. Where is he? I'll get my coat!"

Barbie heard it all come out in a panicked rush. She watched as her mother hurried to the coat closet in the kitchen.

"Ellen, I've got bad news," Tom said. He followed her into the room.

Barbie said nothing, too afraid to speak. She kept her eyes on her mother's face. "A sheriff called from up in Maine," he said. "There was an accident. The boat was caught in this storm. Jimmy didn't make it."

"It's a mistake," Ellie said, shaking her head.

"When the dispatch came in," said Tom, "I asked if I could come tell you."

Ellie turned then and, eyes closed, slammed her head against the refrigerator door, again and again. "Jimmy! Jimmy! Jimmy!" she kept repeating.

"Mama," Barbie cried. "Mama, please stop."

The policeman moved closer and touched Ellie's arm, hoping to calm her.

"He needs to come home!" she shouted.

"There's a blizzard," said Tom. "It'll be a day, maybe two."

"I want Jimmy home!" Ellie cried again. "Do you hear me, Tommy? I want my husband home!"

She was repeating the words as Barbie ran to the telephone and called Bubbles, her mother's best friend.

THE SAILOR IN BAR HARBOR

As moviegoers filed out the front door of the Criterion, it was snowing hard enough that Paul Delaney should have been concerned. But he was twenty-one years old and Mona was pretty. There was the usual excited chatter that follows a film and so they had walked down Cottage Street to Harris's Soda Shop. Mona was more impressed than ever with Lotta Crabtree. Paul doubted much of what he had seen was actually true, and he hoped he never had to sit through another musical. But so long as his date was happy, he was fine.

Snowflakes were falling and three fresh inches had accumulated on the sidewalk. Maybe he should have worn more than a thin topcoat. And he had meant to put his overshoes in the car, just in case. While the situation of the storm concerned him, he didn't see how a couple more hours in Bar Harbor would matter. Already a town snowplow was scraping noisily down the street. Paul took that as a good sign that the roads would be cleared in other areas along Route 1. He had driven in snowstorms before without any problems.

Inside the warm café, surrounded by pleasant aromas and patrons talking excitedly about the movie, a blizzard seemed impossible. He and Mona placed their orders for food and settled into one of the comfortable booths.

THE ICE FISHERMEN

The early dusk of February had already descended. Jimmy Morrill and Pete Godley decided that fishing longer would be playing with fate. And they'd had enough of that in their lives. Besides, they were getting hungry. They'd long ago eaten the last of the sandwiches and donuts. The bonfire that had lit the area in a serene orange glow was slowly dying down as they had planned. It was time to go home. Before they could take their lines from the water a wall of snow and wind came out of nowhere, as if unleashed by a giant fan. It swept in gusts down over the Narrows. The cousins saw it coming, as though it were the shadow of an eclipse, blocking out everything below.

"Holy smokes," said Jimmy. "That's a squall."

Pete helped him get the tip-ups and lines out of the water and into the trunk of the car. Although the snow was heavy and wet, they would be fine. The squall would pass over in a couple minutes. They had four miles to drive back up the lake. But visibility was now limited. With Laddie excited in the backseat, Pete started the engine and flicked on the headlights. Turning the steering wheel slowly, the tires churning in snow, he pointed the

Hillman in the direction of Hanson's Landing. The car crept ahead as he tried to retrace their tracks from that morning. But within a quarter mile, they had already disappeared in snow.

"Just keep the steering wheel straight," said Jimmy. "At least we're headed in the right direction."

What they didn't expect was how much snow the wind had blown onto the upper half of the lake while they were fishing. They managed to drive ahead slowly for almost two miles until the drifts stopped them. The car was now mired in snow.

"Do we try walking?" asked Jimmy.

"It's either that or sleep in the car," said Pete."

It struck them at first as amusing. Two seasoned WWII veterans held prisoner on a lake in Maine.

"It's going to be freezing tonight," Pete noted.

"We'll share Laddie," Jimmy offered. "He's like a little stove."

Walking out would mean coming back for the Hillman when the storm was over. But walking also meant over two miles to Hanson's Landing where Camp Jordan, a summer retreat for youngsters, had been built on the lake. The camp was closed but caretakers would be there. There would be a telephone, and it would eliminate battling snow another mile to the highway.

"Let's see what we're up against," Jimmy said.

They got out of the car and waded in snow a few feet to survey the situation. Laddie followed in their tracks. In no time the snow came to their knees. Now the wind blew so fiercely it was impossible to discern direction. There was no shoreline to be seen, just a blanket of stinging snow. The lake itself was surrounded by hundreds of acres of trees. The only access to it was Hanson's Road, which now seemed like a mouse hole. If they trudged a hundred more feet would they remain headed toward Hanson's Landing and Camp Jordan? Or would they become disoriented and end up lost on the lake?

"We'd be taking a chance," said Jimmy. "And it's almost dark."

The English-born Pete Godley certainly knew snow. But not snow like this. He had never seen so much snow in his life until he arrived in the Pine

Tree State. His original plans after the war were not to leave England for Maine. Anything but. He had his heart set on New Zealand. Those Aussie soldiers he had fought beside in North Africa had convinced him. "It's God's country, Pete," they told him often. "We've been there. We've seen it with our own eyes." And in the midst of the blowing desert sands, the poisonous vipers, the swarms of biting insects, the hunger, the agonizing heat, nothing had ever sounded so good. He never forgot New Zealand, that paradise afloat in the South Pacific with breathtaking fiords, rolling green hills, and majestic mountains. So what the hell was he doing standing in the middle of a lake in Maine, in a blizzard? It was a long story.

"Something to consider," said Pete. "If we run the heater, we'll probably be out of gas by midnight. It'll be a wild night."

They had left Brewer at dawn and before the local service stations were open. And then it was a Sunday. Peter figured a quarter tank was more than enough for the round trip. But just in case, he had plans to stop at the grocery store a couple miles from Hanson's Road on their way home. It had a gas pump. Lesson learned. Even idling the engine to run the heater in increments would be costly.

Jimmy Morrill, like most Mainers, knew the rules. Unless you have a better option a reasonable distance away, preferably a house, you stay put in a blizzard. Car or no car. You wait for conditions to change, or for a rescue. Distances are distorted in falling and blowing snow. And there can be lurking dangers beneath the surface, holes or hazardous objects. Jimmy also knew those rules were often broken, depending on circumstances. It wasn't like they were marooned on an ice floe in the Arctic.

Pete had also learned about staying put during those two and a half years he spent in Africa, waiting for the relentless wind to cease blowing clouds of sand in their faces and wondering what the hell Rommel was up to. Desert sand and heat had done in the engines of many Jeeps and tanks. If you can't see the road ahead of you, don't take it.

"They'll be worried about us," Jimmy said, thinking of family back in Brewer.

"So what do you think?" Pete asked.

Both men knew the seriousness of their situation. The temperature was now dipping lower and a frigid night was riding in on the storm. Should they try their luck? They had been lucky as hell in the military. But maybe their good fortune had run out and they didn't know it. If it had, they would be trapped in the elements all night, at the mercy of a northeaster.

THE LOCAL GAME WARDEN

It was 6:30 and Raymond Morse, a game warden supervisor in Ellsworth, was late sitting down to a warm supper. Morse had spent a grueling day patrolling his area. It looked like the storm was going to pay them a serious visit. How serious he still didn't know. The weathermen had missed calling it and radio reports were vague. He figured from experience it wouldn't be a quiet Sunday night. When the phone rang, he knew he was right. A resident was calling to report that as he crested the hill on Route 1A near Branch Lake, he could see over the tree line and down to the Narrows. The headlights of a car were barely noticeable in the storm. Could someone be in trouble down there on the lake, maybe a fisherman?

Morse was a big man, six feet tall with brown eyes and graying brown hair. On his left forearm was a tattooed anchor from his high school days. A seasoned professional at fifty-one years old, the warden was used to the unexpected when it came to his job. Covering miles of territory and raising families on meager salaries, he and other wardens used their personal cars for travel and bought many needed items from their own paychecks. Morse had even purchased an outboard motor for his rowboat. Otherwise, fishermen would leave him in their wakes before he could check for a license or an illegal catch. Wardens were famous for being overworked and underpaid. A telephone call that someone might be stranded on Branch Lake in a blizzard was run of the mill.

Ray hung up the phone and drove the ten miles northwest on 1A. His windshield wipers battled icy snow that was now being driven by gale winds, the worst kind of situation for people and animals. It was impossible in that visibility to see down on the Narrows, which is what locals called the shoal area midway the lake. When he reached Hanson's Road he turned in, but a foot of snow stopped him. The gravel road was impassable by automobile. Grabbing a flashlight, he followed a set of car tracks that were visible in places where the corridor of trees had sheltered the road. But he couldn't tell if the tracks were coming or going. Maybe whoever had been out there on the lake had driven off and was now home having his own warm supper.

Ray waded through snow back to his car. On the drive home to Ellsworth he met one of the town plows chugging along. If the snow kept falling at this rate it would be a busy night.

THE ICE FOLLIES FANS

It had been apparent since midafternoon that cities and towns in the lower part of Maine were in for one hell of a snowstorm, maybe the worst they'd seen in years. Since the first flakes had fallen that morning gravel trucks and work crews had been hauling sand and salt to slippery intersections and hillsides where accidents were known to happen. With the buildup mounting by midafternoon, plow trucks were put into action. Many V-plows and angle-plows went to work in a dozen counties clearing streets and highways, the larger ones breaking the track and the smaller plows following to widen it.

But the snow was getting heavier and the winds picking up power from off the ocean. The machines were able to handle the accumulation at first, but it became difficult to sustain the pace. Men on the work crews soon understood they'd be at this grind for the rest of the night and possibly all the next day. And travelers on the road, especially on coastal Route 1 from Eastport down to Kittery, were now realizing their own predicaments—as did motorists on the short stretch of new turnpike from Kittery to South Portland. These main arteries were becoming impassable. Concern was now growing among family and friends of the Ice Follies fans who had driven from all over Maine down to Boston Garden to attend the afternoon matinee.

Stanley Peterson was the driver of the chartered bus that left Bath that morning as Bill Dwyer was buying his Sunday paper. Always dependable, Peterson had telephoned the owner of the Bath Bus Service to report that he was leaving Boston with his passengers. That was at 6:30 P.M. It would take some maneuvering to get out of the city, especially in the procession of automobiles and buses from around New England that were inching from the parking area at the arena. But at least the Bath bus was headed north to Maine with over thirty excited fans aboard. The matinee had been worth the $10 ticket, which included the bus ride. Now they were eager to get home.

CHARLES E. VOYER

Also on the road that night was a lone driver named Charles Voyer, a retired employee at Portland's well-known State Theater on Congress Street. At sixty-three years old and with a heart condition, Voyer had hesitated making the drive down to Boston by himself. But Harriette, his second wife of just five years, was not feeling well. Voyer had spent his career in stage production, involved in everything from scenery and props to rigging and lighting. His first theater job as a young man was working at the impressive Music Hall in Lewiston, home for decades to the best opera and stock companies in the country. A chance to witness the elaborate production put on by the Ice Follies was worth a drive in what was forecast to be unpleasant but manageable weather.

Charles Voyer left a half hour earlier than the chartered buses from around New England that had lined up at the exit gates waiting to load their passengers. When he reached the Maine state line, he would take the new turnpike from Kittery to its end at South Portland. This exit was not far from his and Harriette's home on Cumberland Avenue. By the time Voyer drove his 1947

Dodge sedan past the Welcome to Maine sign it was almost eight o'clock and the storm had greatly intensified. Cars inched along in the snow, grateful for the guiding red taillights in front of them. And thankful that large trucks and buses were paving the way. On the turnpike, the snow was now eight inches deep.

DELANEY LEAVES BAR HARBOR

Harris's had closed and the other patrons disappeared by the time the infatu-ated young sailor from New York realized he should get on the road. Mona had been encouraging him to do so ever since the movie ended. It wasn't a big deal, he assured her. He would be fine. Winter Harbor wasn't far, and the plows would take care of the storm as they always did in rural Maine. Stepping out onto the street, Paul Delaney saw three more inches of fresh snow on the sidewalk. Maybe in showing off his bravery he had been more smitten than wise. He thought of kissing Mona goodbye and then, reading her reserve, decided to wait until next time. She had already agreed to a second date when he had his next weekend pass. On impulse, Paul gave her a quick goodbye hug that she didn't seem to mind. He lit a cigarette and watched her walk up Cottage Street, her boots trailing through the snow until she turned the corner.

In the parking lot near the theater Paul brushed the five inches of snow that had collected on the windshield since he parked the car. He got behind the wheel and edged out of the parking lot. He turned onto Eden Street, following what is a spectacular view of the ocean in good weather. All he saw then were waves of snow beating at the car. Along that road, the greatest mansions in the world had once sprawled. Some were still there. With his wipers working furiously, the snow blinding him, he didn't notice the well-lit villa to his right, towering over the ocean. Off-season, it was likely empty of rich and famous guests, with just a household staff doing upkeep. Its name was Buonriposo, once the summer home of Ernesto and Edith Fabbri. It was on those magnificent grounds that Alessandro Fabbri, in order to prove his loyalty to the United States, had conceived of the receiving station at Otter Bank that Rockefeller later moved to Winter Harbor, the same station Paul had driven away from that afternoon.

Less than three miles past the magnificent Fabbri "cottage," Paul reached the foot of Ireson Hill. He had realized his folly back when he saw snow piled on the windshield. Mona was right. He should have departed earlier. He pushed gently on the accelerator pedal and began the climb that would curve up and around near Hulls Cove. Halfway up the hill, the car fishtailed. Panicked, Paul hit the brakes. Buoyed by the wet snow that was freezing on the hillside, there was no stopping the vehicle. He sat helpless behind the wheel as it slid backward. At the bottom of the hill the car skidded, spinning out into a ditch filled with snow. His heart thumping, hands shaking, at first he couldn't move. Where was the ocean? Visibility had already been bad on his drive in. Was it anywhere near the road? Would he topple over into the water and disappear?

Calmer now, he pushed open the driver's door and got out. All around him snow billowed in gusts. He tried to shield his eyes so that he could assess the situation. There was no way he could back the car out of that snow-filled ditch. Because of the pounding storm, he didn't see the lights of a residence sitting a hundred and fifty feet away. Just as he hadn't seen the sign at the foot of the hill that warned DETOUR. Even the locals didn't want to challenge Ireson Hill in a snowstorm.

Paul waded back to the car, his only refuge in that tempest. Sailors had been trained when they arrived at the Maine station that it was wise to carry supplies in one's vehicle during the winter months in case they became stranded in a snowstorm. But no one paid much attention to that warning. Paul had brought two bottles of orange soda with him. He already drank one on the drive over and intended to drink the other on the return. But the plow would be along any time and would stop to pull him out. He wished now he'd bought a candy bar after the movie. His good luck was that Jeffrey had thrown two blankets into the back seat. It was already getting cold in the car. He could run the engine but he wanted to save fuel for the trip to Winter Harbor. At least he had bought that pack of Chesterfields at the theater. He lit one and smoked it with the narrow vent window pushed open. When he finished, he flicked the butt out into the snow and closed the window. He lay beneath the blankets to wait. In minutes he dozed off.

SUNDAY EVENING

The snow had begun in the gloaming.
And busily all the night
Had been heaping field and highway
With a silence deep and white.

Every pine and fir and hemlock
Were ermine too dear for an earl,
And the poorest twig on the elm-tree
Was ridged inch-deep with pearl.

—"The First Snowfall" by James Russell Lowell

NINE O'CLOCK IN THE TOWN OF BATH

In downtown Bath plow trucks were still struggling to keep the streets cleared. Otherwise, emergency calls for policemen or ambulances would be placed in vain. Knowing that a firetruck might have difficulty reaching residents with a house fire, the local radio stations made public announcements urging people to be careful when heating their homes. They should fill bathtubs, wooden buckets, and tin pails with water just in case it was needed. Those living near fire hydrants were required to keep them shoveled out during the storm. It was now obvious that this northeaster wasn't neglecting southern and coastal Maine. In fact, it seemed to have those parts of the state in its crosshairs.

The Tardiffs, like other Maine families accustomed to snowstorms, were not yet alarmed. The city had always been reliable when it came to plowing Varney Mill Road, their only route into town. Instead, they had enjoyed a quiet Sunday. Phil served his spaghetti and meatballs for the midday meal. Supper that evening was just his crepes, smothered in brown sugar and butter. Hazel appreciated her husband cooking more than usual that weekend. She hadn't slept comfortably the past few nights, and there was swelling in her legs. Even the kids were on their best behavior. After supper, the girls had done the dishes without being asked. And David helped his father shovel a foot of snow from around the Buick. But by the time they cleared the driveway leading down to the road, wind coming off the Kennebec River had covered their hard work with more snow.

"We'll shovel again in the morning," Phil said. In the gleam of the porch light, he stared out at the road. He hoped Hazel was right, that the baby wouldn't come that night. Only the city's big Walter plow, weighing a few tons, would be able to clear that road now. He said nothing to David as they kicked snow from their boots at the kitchen door.

"Did the plow make it out?" Hazel asked.

"Not yet," said Phil. His hands now warmed, he lit a cigarette. "But it will."

Phil had work in the morning. It took a major storm before Bath Iron Works shut down its operation. There was no doubt that school would be canceled. Hazel would call the principal's office in the morning if the local radio station didn't announce closings.

As Phil and the kids listened to *Inner Sanctum* on the small tube radio in the living room, Hazel went to the kitchen to make sandwiches. The kids took a bagged lunch to school each day, usually peanut butter and jelly sandwiches. If school was canceled in the morning, and it likely would be, they could eat them for dinner. Into Phil's metal lunchbox she put two baked bean sandwiches, his favorite. That Saturday night pot of beans was never wasted. The kids wouldn't eat bean sandwiches but Phil loved them. Hazel cut thick slices of her homemade bread and covered the beans with a layer of ketchup. In the morning, she would make a pot of coffee to fill the thermos that fit snugly into the lunchbox's rounded top.

When Phil and the children had gone to bed, Hazel lingered in the kitchen to smoke a bedtime cigarette. A couple of her nurse friends and her mother May, who had also been a nurse, were opposed to smoking while pregnant. But no doctors that Hazel knew personally had warned her of any dangers. Yet, she noticed how the kids often coughed and waved away the smoke when their parents lit up cigarettes. Hazel had cut back on her Pall Malls until the baby came. And she promised her children yet again that she would quit one day soon.

With wind whistling snow around the windows, she wiped bread crumbs from the counter and rinsed the butter knife in the sink. The girls were old enough to make their own lunches, but Hazel liked doing it. She had given up nursing school to become a housewife and mother. There were some regrets. She hadn't forgotten her dreams while growing up on Isle au Haut. But hearing the children's voices earlier from the living room made all the difference. She wasn't worried about the baby coming, but she had concerns. It was the unknown that made Hazel apprehensive. There were times when she woke from an unsettling dream, certain she heard a hammer tapping nails into a small coffin. Her parents had gone through the loss of their stillborn child on that cold December day in 1914. But that was a long time ago, that other life the Coombs family had lived on Isle au Haut, a place far from hospitals and doctors, a small plot of land jutting from the ocean.

Before she went to bed, Hazel Tardiff stared out the window at the swirling snow, laden with river wind. The whole world seemed buried in white. Then she turned out the porch light.

GEORGE ASPEY OF WARREN

George Albert Aspey, known as "Bert" to his friends and fellow workers, stood in the house he shared with his mother and stared down the embankment that led from their front steps to Main Street. It was one hell of a storm, so much so that George couldn't see the hundred-foot-tall chimney rising up from the Georges River Woolen Mill in the distance. Snow swirled in clouds down toward the town, wind whipping it around automobiles that sat unmoving along the street. He wondered if the mill would cancel work the next day. Maine factories and businesses were tough when it came to snow days. If they weren't, they'd shut down too often to make a profit during the winters.

"How does it look?" Annie asked from her chair in the parlor.

"Not good, Mother," said George.

For over a hundred years a woolen mill had sprawled on the banks of the Saint George River in the town of Warren. Locals were proud of their mill. The chimney rising up from the boiler room was erected using 100,000 bricks. While the peg mill in Brownville had once cranked out shoe shanks for the military, and Bath Iron Works built those sturdy ships, the woolen mill in Warren made blankets for Uncle Sam during World Wars I and II. George Aspey had contributed to the making of those blankets in the second war. Now he was employed there on the first floor as a carder.

"I hope Hilda will find a boy to shovel," said Annie.

"If the mill closes tomorrow, I can go," said George.

Hilda was his only sister, five years younger. A music lover, she had once been a weaver on the second floor of the mill. She was now a housekeeper for a family who lived out on Rod Road in the summers and escaped to warmer climes during winters. Hilda was proof of the Aspey family's solid roots in a new country. Only she had been born in Maine. George, his older brother Harry, and their parents had been born in Northwich, England. In the summer of 1901, when George was five years old, his family had immigrated to the United States, arriving at Portland from Liverpool aboard the SS *Vancouver*. As with shipbuilding in Bath, and with those quarries and railroads that had helped build Brownville, the mill in Warren had employed generations of local families. Over the years, all of the Aspeys at one time or another found work in the mill. Now, with his father long deceased and Harry having put down roots in Boston, George shared a house with his elderly mother.

At fifty-six years old, George Aspey stood five-foot-six. His blue eyes were accented by abundant gray-white hair that had been blond in his younger years. He had readily served his new country in World War I, becoming a naturalized citizen after he was inducted in late 1917. The war over, he came home to find his previous job as a carder at the mill had been filled. Seeking work, George ended up living in a rented room in Bath and working as a chipper at Bath Iron Works. Bill Dwyer, newly married to Nellie, was a riveter there at the time. But Aspey missed his family back in Warren. When a job opened at the woolen mill, he returned there to live out his life. A personable man who never married, he was well-liked in town, active at church and in several organizations. Over the years, the masonic lodge in Warren had elected him to serve in several high-ranking positions.

When his mother retired for the night, George opened the front door and stepped out on the porch. On summer and autumn Sundays, he and Annie enjoyed sitting out there in comfortable chairs after church. Sometimes, if the choir practiced that day, they heard sweet voices rising up from below, Hilda's being one of the strongest altos. You could see the woolen mill from the porch when the leaves had fallen. George considered the quarter-mile walk to his job as good for his health. It was even a pleasant walk on most days, except in downpours of rain. He would stride down Riverside to where it met Main at the bottom of the hill. Turning left, the bridge would carry him over the Saint George River. He would pass the masonic hall and then the cafe where mill workers could order the best homemade pie in Maine. If he instead turned right on Main Street, he could walk past Simmons Funeral Home and on up the steep hill to Rod Road where Hilda now lived.

But on this Sunday night the porch was buried in snow, and cold wind beat at George's shirt. Under the streetlight, he saw a foot of snow filling Riverside with more piled higher at the bottom. Down on Main Street the blurry yellow headlights of a snowplow struggled in the drifts. George closed the door and left his work boots on the rug where he would find them in the morning.

RAY "SONNY" POMELOW, IN BROWNVILLE

"It's like being inside a snow globe," Sonny said. He shook the melted flakes from his coat and then hung it on a nail by the kitchen door. "It's really coming down." He had just finished shoveling the front porch.

"There won't be any school tomorrow," said Grace.

Sonny figured his mother was right, the way the stuff was falling. He didn't say it, but how could a guy not be happy if school was canceled? He knew Johnny, Bobby, and his other friends were hoping for the same. No school. That it was a Monday made it even sweeter. Making up for a snow day in warmer months wasn't fun, but summer was a long way off.

"I wish we could've gone to Gordon's party," said Sonny. His sister had expected them to leave before noon. And then the snow started falling. Etna was fifty miles south over a narrow, two-lane highway.

"Louise didn't want us coming in this weather," said Grace. It was her grandson's birthday and Louise had planned a small party. Turning five years old, Gordon was the little brother Sonny had always wanted. He hoped one day soon that his sister and her husband would find work in Brownville and move back home. He took off the mittens Louise had knitted for him and threw them on the register. The balls of frozen snow would soon be sizzling into droplets and then gone, leaving behind that wet winter smell that all Mainers knew well. He pulled off his rubber boots. Next to the wet mittens he lay one of his woolen socks.

"I got a boot that's leaking," said Sonny. "My sock is all wet."

"Dad will help you patch it tomorrow," said Grace. "Them boots need to last you a while longer."

Sonny nodded. He would get out the repair kit that Ray Sr. used to fix car tires. Sonny used it for his bike tires, too. It was a cardboard can with a metal top and bottom that held a tube of glue and a swatch of rubber to cut patches from. On the can a camel was standing in front of a tiny pyramid and a couple of scrawny palm trees. The word CAMEL was in red letters. The metal top had ridges, like a built-in file, to sand down the area that needed repair. There wasn't a working family in Maine that didn't have that can sitting somewhere on a shelf. Ray Sr. once remarked that the can should have a moose on it, not a camel.

"You hungry?" Grace asked. When Sonny nodded, she turned a stove burner on under the pan of corn chowder she had made for supper. Sonny got the jar of relief peanut butter down from the cupboard. He would make a peanut and jelly sandwich before he wandered up to bed.

TEN O'CLOCK IN DOWNTOWN BATH

Bill Dwyer had fallen asleep while listening to music on his radio. He hadn't thought of his first wife, Gladys Adams, in much of the forty years since they had parted ways. His life since then had been mostly with Nellie. And yet he had dreamed of Gladys as he slept. She was from Portland and still eighteen when they married. He was almost a decade older and a riveter at Bath Iron Works. They were wed one warm summer evening in the parsonage of the local Baptist church, Gladys wearing a white dress trimmed in blue. After a small reception in his mother's parlor the newlyweds left on the midnight train to Manchester, New Hampshire, and then down to Boston. Gladys looked like she had stepped from one of those fashion magazines she loved, her going-away dress a blue serge and her white hat sporting ostrich feathers. "They're willow plumes, Bill," she said proudly when she first showed him the hat. "They're more special." They had nothing but future ahead of them that night as they boarded the train. Who knows when or why love goes away. Eight years later he was married to Nellie.

The clock's hands showed a few minutes after ten. Bill usually let Snooky in an hour earlier than that. "Us two old bachelors go to bed early," he had once told a neighbor. It was snowing hard when he opened his front door. He expected to see the cat waiting there, his coat wet with flakes before he padded inside to his bowl in the kitchen. Wind had blown six inches of fresh snow onto the porch. And in the front yard there looked to be well over a foot of the stuff. Snow gusted in sheets and drifted out into the street. It was impossible to see beyond his own yard. Where was that damn cat? Bill noticed paw prints around the front door, now almost obliterated with snow. So Snooky had returned, waited, and then decided that his owner had likely dozed off. He had gone back into the neighborhood for a final prowl.

When he finished his bedtime cocoa and toast, Bill opened the door again, hoping to hear an impatient *meow* greeting him. Snooky was not the kind of cat to suffer fool owners gladly. But there was no sign of him. Snow blew off the rooftop and swooped down into the front yard. Trees lining the other side of the street swayed in the wind. Wind *and* snow. That was not a harmonious pair. He hoped the residents of Bath were off the roads at a time like this and safe at home. Likely, the plows and work crews would be out all night. If a fire started at some house, it would mean trouble on those clogged streets. He knew this from experience, from his years with Volunteer Engine No. 4. Again he felt that pang of nostalgia. It had been gratifying to know that he was an asset to his community.

He called to the cat a few times, waiting for a reply as he stared out into the night. There were sheds and porches in the neighborhood where a smart cat could sit out a storm. He watched as snow curled like smoke around his front door and sprays blew past him into the living room. He would check again later when he rose in the middle of the night to use the bathroom.

"Ostrich feathers on a hat," Bill thought as he closed the door. "Whoever came up with such a foolish notion?"

HULLS COVE IN BAR HARBOR

When Paul Delaney woke, he didn't know where he was. Only that he was shivering, and it was dark in the car. Then he remembered. His car had slid down Ireson Hill and off into a side ditch. He had been waiting for the plow. Had it come and gone while he slept? He rubbed his eyes and sat up behind the wheel. How had it gotten so dark? He pulled his cigarette lighter from his coat pocket and flicked it on. His watch read half past eleven. He held the lighter up like a torch so he could see in the car. The windshield was solid white. The side windows were all white. The rear windshield was white. "Snow must have drifted over the car while I slept," he thought. He would get out and brush it off, then see if there were any signs of the plow.

The driver's door would not move more than a quarter inch. Paul tried turning the steering wheel to loosen the tires. They didn't move. He leaned over and tried to open the passenger door. It was blocked. He crawled into the backseat. Even using his feet to push on the doors, neither would open more than that small space. How could it be possible that the car was buried in snow while he slept? But it was. He talked to himself to keep calm.

"This is just some bad luck, Paul," he said. "The plow will see you."

But how? If the snow covering him was deep, not even his brake lights would be visible. He turned the key in the ignition and the engine started. He felt a wave of relief wash over him. At least he could stay warm. He knew not to run the car for than a few minutes at a time, something else the young sailors to Maine had been taught. But the exhaust pipe must be clogged with snow if the entire car was covered. That risked carbon monoxide poisoning. If only he had dressed more warmly. At least he had those two blankets his roommate had wisely put in the car. His toes had grown cold as he slept. But who would wear dress shoes if they knew they'd be caught in a blizzard? He had left Winter Harbor that day to take a pretty girl to the movies, not spend the night in a ditch covered with snow.

Feeling the need to urinate, he thought of the empty pop bottle on the floor of the car. But he might need it later to fill with snow for drinking water. For now, he would use the narrow space where the back door opened into the snow, and do his best to aim well. Surely this would all be over soon. He and the guys would have a good laugh back at the base. He could hear them now, teasing him about pissing in the snow.

"Keep your cool, Paul," he whispered. He lit a Chesterfield and inhaled it, feeling it calm him. It was his good luck that he just bought that pack and only three were gone.

ALICE AND CAROLYN

Alice checked on her sleeping daughter, making sure she was well covered. Then she went into the tiny living room and sat on the sofa. She felt exhausted from crying, and more so in trying to hide it from Carolyn. Beyond her view, a hundred yards down to the water, she knew Harland's wharf would be buried in snow. Wind rocked the trailer. She tried not to think of how cold the water must have been, how even someone as brave as Harland would have been frightened. Had he thought of them in his last minutes? Of course he had. She hoped it had given him strength when he would need it most.

The small pine bookcase he made by hand sat near the sofa, its two shelves filled with books on boat design and fishing. Alice pulled out the one book that was his prized possession. *My World Is an Island*, signed to Harland by the author Elisabeth Ogilvie. Two of the pages were dog-eared where she had mentioned him.

There's a silvery poplar by the Davis fishhouse, gleaming and twinkling with every breath of wind. The just-painted mast and boom—bright orange—on Harland [sic] *Davis's new wharf stands out brilliantly . . . Everybody used to fish off Walter Young's lobster car; this year, they're fishing off the float by Harlan*[sic] *Davis's new wharf.*

Alice put the book away. One day, she would pass it down to family. And the same with those postcards Harland had mailed her when they were courting. She lived just a few miles up the road at West Waldoboro then. She would never forget the thrill of finding one in her mailbox. He didn't have any writing paper so he had stepped into a Thomaston drugstore and bought a half dozen cards, the kind summer people usually buy. They were of Pleasant Point and a couple pictures showed his wharf at low tide. Harland was pleased to think of the strangers around the country who had received those postcards, even if no one knew whose wharf it was. He was proud of how well his business was doing. She read the words on the back of one card. *That's a new derrick. I broke the old one lifting a millstone. The lobster boats are moored just off the end, about 30 of them. Trucks back out to the derrick. The wharf has held ten tons (truck and loader.) I have let it out to many. Never a dull moment. Why don't you drop me a line about the time? If you are at the theater I will know where to find you.*

They had met up that weekend at the Strand Theater in Rockland as planned. *Sunset Boulevard* was a much-praised film and Alice wanted to see it. Harland hugged her in greeting and then told her that his mother was taking new medicine and feeling better. Always fond of his mother, he was hopeful that day in 1950. Alice knew now the medicine hadn't worked. Eva Creamer Davis was so crippled by arthritis she was mostly bedridden. The sad news about her only son would devastate her.

Alice put the postcards back into the envelope that held them. Something told her she would read those words a million times down the years of her life. *If you are at the theater I will know where to find you.* She looked over at Harland's brown slippers waiting by the front door. And Carolyn's winter coat hanging neatly on the coatrack. She had chided her daughter earlier in the day for tossing it onto the sofa. "Carolyn, don't throw your clothes down like that." What foolish things humans demand of each other, she thought now. Or when Harland came in with boots wet or muddy and didn't stop on the rug to wipe them. She had nagged him about it. He would catch her by the waist saying, "I thought you'd be mad if I got your nice rug dirty," and she would laugh and forget until the next time. The trivial, insignificant things that human beings worry about when all that matters in the world is love.

"Dear heavenly father," Alice said aloud. "This is a heavy grief you have given me. Please forgive me if I fail to measure up."

She had a daughter to raise, and a baby coming in five months.

55 GATES STREET, PORTSMOUTH

Ellie Haigh couldn't sleep. She had finally accepted the fact that her husband was not coming home on that stormy night. Both he and Harland Davis were at a funeral home in Maine, waiting for the employee who did embalmings to struggle through the snow the next morning and begin his work. In her grief, Ellie had asked Earle Sanders to drive the ten miles to Dover where Jimmy's mother and father lived. Ellie knew she couldn't give her in-laws news like that over the phone. And she was in no shape to leave the house that night, snowstorm or not. She tried to think of her future, but it was still hazy, Jimmy's death too new. Still, her thoughts kept coming around to their daughter. How was Barbara Ann going to manage without Jimmy there to guide her? Barbie idolized her father.

Edith Finnigan, known to her friends as Bubbles, had arrived earlier in the evening and taken over, doing what she could to assuage the awful situation. Less than a hundred pounds at under five feet, Bubbles could be a whirlwind when the chips were down. She spent the evening answering the phone and taking messages. She discussed food with friends and relatives, casseroles and sandwiches, so that Ellie wouldn't have that extra worry. She informed the local funeral home that it might be a couple days before they could make arrangements for transporting Mr. Haigh back to Portsmouth. And she finally convinced her good friend to go to bed and try to sleep. Ellie would need her strength for the days to come.

Barbara Ann Haigh had been trying not to cry for her mother's sake. None of it seemed real to her yet. It was more like a bad dream happening in the world of adults. On those rare nights when she had a nightmare, she had always crawled into bed between her parents to feel safe. There was the time when she was barely six years old and her father had taken her to see *Abbott and Costello Meet Frankenstein*, thinking the comedy would make her laugh. Instead, the monster had terrified her. She slept between her parents for two weeks, with Ellie scolding Jimmy for his choice in movies. Where would she go now in the dark of night?

"Barbie, do you want to sleep downstairs with me tonight?" Ellie asked. "I don't want to sleep alone. You can even bring Skybow into the bedroom."

"I can?" Barbie asked. Skybow had never been allowed upstairs in her own bedroom before. And he had certainly not been allowed in her parents' bedroom. For a moment, the sorrow lessened. It's what Ellie wanted, to move her daughter's mind away from the awful truth. Ellie had grown up poor, in a house full of siblings and a lot of kids in one bed. She had learned young that there can be a safety in numbers.

"Bubbles can sleep in your bed," she added.

Barbie stood looking into her parents' bedroom, the neat pillowcases, the folded quilt at the foot of the bed, the family picture of the three of them on the dresser. Skybow sat next to her, waiting, as if uncertain of this windfall. This was the dog who was Jimmy's miracle mutt. As a puppy Skybow had been hit by a speeding car. When the driver didn't stop, Jimmy carried the little dog to the vet and spent a fortune on three surgeries to save him. He was as good as new, but minus one leg that he didn't seem to miss in his boundless energy. Jimmy taught Barbie how to change the dressings until the stump healed. "I had a dog when I was a kid," Jimmy told her, "and I named him Skybow." Barbie smiled. She knew that's what she would call her dog.

Satisfied that he was welcome, Skybow curled on the rug near the bed. Barbie wasn't sure what to do. Should she lie where her father used to sleep? It didn't seem right. She remembered the Sunday mornings, pillows propped behind him, that he drank an orange juice while reading the newspaper in bed. Finally, she crawled under the sheets on her mother's side and waited. She wasn't sure if she imagined it, but she thought she smelled Old Spice cologne on the pillowcase where her father's head had lain that morning. When Ellie came to bed, they held each other.

"We have to learn to live without him, sweetheart," Ellie whispered. "We have no choice. Can you be brave tomorrow when Grammy Haigh gets here? She already lost Uncle William. Maybe, if we think of Grammy first, it will help us carry our own grief." Jimmy's mother was English-born, coming from a little town in West Yorkshire. She was affectionate, but also quite British and reserved about showing one's emotions in public.

"I won't cry, I promise," said Barbie.

"I'm proud of you," Ellie told her. "Daddy would be proud of you, too."

Barbie reached a hand down by the side of the bed and touched Skybow's soft fur.

This is how it's going to be from now on, she told herself. *This is your new life, Barbara Ann. So be a big girl and don't cry.*

THE MORRILLS, IN BREWER

When Jimmy and Peter didn't return home by seven o'clock no one in the Morrill family noticed. They were enjoying a quiet Sunday night at home, as well as preparing for the week that lay ahead. The snow had been slowly piling up on the streets of Bangor and Brewer, but traffic was still manageable and plows were busy clearing the streets. At eight o'clock, remembering the fishermen, Richard looked down from his apartment on the second floor to see if the Hillman was back, parked next to Jimmy's Rambler. It wasn't. He informed his brother Carlton, on the third floor. The Morrill family was not overly concerned. Even disregarding their military experiences, Jimmy and Pete were known as keen outdoorsmen.

At nine o'clock their worry began to escalate, as had the storm. Branch Lake was sitting much closer to the coast than Brewer. Were storm conditions worse there? Had something gone wrong? On the other hand, they were grown men. What if they had run into friends and went for supper at a local diner? And maybe after they ate, they decided on a card game or to shoot a game of pool? There were a couple places open on Sunday night. Or maybe they had car trouble and that was causing the delay. But if any of those things had happened, why hadn't Jimmy or Pete phoned to let the family know? Maybe they should call the police, just in case. The last thing anyone wanted was to report these two experienced veterans as missing when they might walk in the front door at any minute, lugging a pail of fish. "Jumping Jehoshaphat!" Jimmy would likely say, one of his favorite expressions. "Can't a couple of guys go fishing without you folks calling the cops?"

At ten o'clock the Morrills called the Brewer police and reported the pair missing. There was nothing the local police could do. Ellsworth was not their jurisdiction. And there were enough storm emergencies to tend to within city limits. Familiar with Ray Morse, the game warden supervisor over in Ellsworth, the Brewer police chief phoned Morse's home. His wife answered, her voice sleepy. Ray wasn't there. He had gotten a call from a friend of his, a farmer on the outskirts of town who needed help getting his livestock into the barn before the storm got worse. He hadn't returned yet. The chief left a message to give her husband when he got home. Two Brewer men hadn't returned from ice fishing on Branch Lake. Family members were getting ready to drive down there themselves and start searching. They could use the game warden's help when he returned.

Carlton then phoned two family friends the Morrills could count on to help. Louis Stearns and Foster Johnson both knew Branch Lake well having fished it before with Jimmy and Pete. They promised to get to the tavern as fast as they could. As the brothers got out their snowshoes and found flashlights, Cecile Morrill, the French-speaking girl who had married Richard, began making sandwiches from the roast beef left over from supper. Stella, married to Carlton, boiled hot coffee enough to fill four thermoses. The guys might be gone for the entire night. And then, what had Jimmy and Pete eaten once the lunch they had taken that morning was gone? Stella packed donuts and Cecile found a pork chop left over from the day before and wrapped it in newspaper for Laddie.

Neither brother spoke as they loaded snowshoes into the trunk of the car. Louis and Foster arrived and added their own gear. The hope was that the plows were out and the road to Ellsworth still open. Over a foot of snow had fallen and more was coming down. It was a slow drive. At times the snow blew at them in sheets, forcing Carlton to pull to the side of the road and wait until the wind died down. It was almost midnight when they saw the sign that said BRANCH LAKE.

PLOWS TO THE RESCUE

By ten o'clock so much snow had fallen that city officials in Bath realized it was time to bring out the big guns. The residents out on Varney Mill Road would have been reassured to learn that the monster plow truck, a 3.5-ton Walter, was being sent to keep their road open for fire and emergencies. The Walter was so huge it lumbered slowly along. It had no heater or defroster. Heat rose to the cab from the manifold below, through cracks in the wooden floorboards. Most models had a top speed of twenty-eight miles an hour and ate a gallon of gas every five miles traveled. At Lover's Retreat, the Walter ran into a ditch when gusting snows blinded the driver. The drift there, thanks to a steadily blowing wind, had already grown to over eight feet deep. It was now impossible to get the Walter back on the road. Hoping to find a telephone, the driver left his assistant in the truck and struggled two miles through yard-deep snow. Icy wind stinging his eyes, he finally glimpsed the yellow lights of a farmhouse. They happened to have a telephone. It was after midnight when he called his highway superintendent to report the accident.

Since there was no doubt it would take a giant of a truck to get the Walter out of a snowbank, the superintendent sent the city's 5.5-ton Walter to the rescue. It rumbled out Brunswick Road, crunching down snow as it went. The windblown drifts were already eight to ten feet deep in places and the storm was still young. At Dunton's Corner the big beast ran into trouble. So much freezing snow had been driven in under its hood that the ignition coated over with ice. This bigger Walter now rolled to a stop, snow piling up around it.

Getting another distress call from this latest driver, the weary superintendent sent out a wrecker, hoping it would be able to free the smaller Walter and get it back on the road. The wrecker got as far as the hill on the old Bath-Brunswick Road when it came upon a state plow that had stalled there. Stopping to assist the plow, the wrecker ended up nose deep in yet another eight-foot-high drift. With all three machines now incapacitated by snow, the distraught superintendent dispatched the city's caterpillar, its sand-and-snow loader, to the rescue. At the bottom of Foote's Hill, the caterpillar ran into a high bank of snow and became stuck there.

The Tardiffs and a dozen other families out on Varney Mill Road, likely all sound asleep, had no idea they were out of luck for the time being.

GAME WARDENS ON THE LAKE

When Ray Morse got back to his house at 1:00 A.M. he was exhausted from slogging around in the snow, helping a farmer round up his cows and get them into the barn. It wasn't a part of his job, but they had been friends for years. The last thing the police or game wardens wanted was for cars to start hitting farm animals they couldn't see on the road. As it was, accidents were happening all over town. None were fatal, but they were serious enough to keep the police busy. Despite some plows in the Ellsworth area making progress in keeping up to the snowfall, there was no doubt in Ray's mind that Route 1 would eventually be shut down. It was getting too dangerous on that narrow road. When he sat down to eat the sandwich his wife had left for him, he saw her note on the table. Two fishermen from Brewer hadn't come home from Branch Lake. Was it their headlights the caller earlier had seen near the Narrows? Judging from when the call came in, the Morrill group had likely reached Branch Lake an hour earlier.

Ray called Robert Hogan, one of his game wardens in Ellsworth. Hogan had also been out most of the night, helping pull cars out of ditches.

"Bob, we got two fishermen somewhere out on the lake," Ray told him.

"Pick me up," Hogan said. "I'll be ready."

Ray filled his thermos with the last of the coffee from the pot on the stove. He would eat the sandwich on the drive up to the lake. He grabbed a warmer coat and an extra flashlight. Everything else he needed was in his car. But the intensity of the storm was not going to make it easy. He figured they should start searching at the upper end of the lake. Ice fishermen on Branch often winter-fished for trout or salmon around those two little islands. But inlets, islands, and narrows were also the places where an automobile *could* break through the ice. The water moving in those areas was slightly warmer and gradually caused ice erosion. It would likely be a long night.

When Ray Morse and Bob Hogan arrived at the turnoff to Hanson's Road, they saw a car parked on the side of Route 1A. The snow around it and leading down toward the lake showed numerous snowshoe prints.

"Those guys from Brewer made it," said Ray. He pulled in front of the car and parked. By the time he got out, Bob was already strapping on snowshoes.

TANGLED TRAFFIC

Stopping had not been an option for those zealous drivers on the Maine Turnpike who felt they could overcome the odds and make it home. In the beginning, traffic was moving at ten and fifteen miles an hour and had been since four o'clock that afternoon. Those who were not Ice Follies fans left before the heaviest snows fell and made it through to the last exit at South Portland. But later motorists were not so lucky. Most state and town plows had been rendered useless. Now hundreds of red taillights braked behind each other as they crawled ahead like metallic ants. After several hours of advancing just twenty or thirty miles, the storm finally got the better of them. The string of stranded vehicles from Kittery north now stretched over the forty-five miles to South Portland. Many sought out nearby homes or places of business to sit out the storm.

Travelers were now faced with the havoc that a determined northeaster could inflict on a targeted area. Winds were lashing the New England coastline at thirty to sixty miles per hour and creating waves at some ports that reached nearly seventy feet. By midnight Maine highway officials decided to shut down the turnpike. State and local police stationed themselves at the toll bridge on the Maine–New Hampshire line. All turnpike traffic headed into the state of Maine was denied entry. Other drivers, at their own risk, decided to try their luck on Route 1.

THE THEATER TECHNICIAN

Charles Voyer had thought he would be home with Harriette in South Portland a few hours earlier. Instead, he had crept along at a snail's pace, following in the tire tracks of trailer trucks and passing more ditched automobiles and ensnared plows than he could count. Voyer had been on the road for nearly seven hours and had advanced ninety miles. It was now almost midnight. Wind swept snow onto the windshield as his wipers fought to clear it away. He had considered taking an exit, but with plows struggling to clear the turnpike and Route 1, the icy snow had piled even higher on roads leading into towns along the way.

Finally, at the Saco exit, fourteen miles from the one that would deliver him safely home, his and other vehicles could fight the drifts no longer. He turned the wheel of his Dodge and edged the car over to the side of the road, praying no bus or plow truck would ram into him. Through the front windshield he saw dozens of blurry red taillights. Through the back windshield a string of yellow headlights. Nothing was moving for forty-five miles of turnpike. All along the toll road hapless motorists hunkered down for a cold night that would bring a cold dawn.

Voyer couldn't believe his luck. This was not the first snowstorm that had waylaid his entertainment plans. When the 1940 Valentine's Day blizzard had struck New England, killing nearly seventy people before it was over, he had taken a train down to the Boston Garden to see Sonja Henie, three-time world champion figure skater turned movie actress. The Weather Bureau had predicted light rain and snow for that day, too. When trains were canceled because of the storm—the snow fell at three inches an hour—he and hundreds of others slept in North Station, huddled on benches in the cold. The luckier ones could afford pricey hotel rooms. Many dozed at reduced rates on cots in hotel lobbies and ballrooms. The local bars were filled with Henie fans. Crews worked through the night with shovels and pickaxes, by the light of bonfires and oil lamps, to free the train tracks of snow. The next morning, the Yankee Flyer had carried Charles and other fans back to Maine.

Now another storm. And this time he would spend the night in his car, while his wife and warm bed waited just miles up the road. Charles undid the extra set of rosary beads that had been dangling for years from his rearview mirror. After he said his prayers, he made the sign of the cross.

THE HOWARD JOHNSON'S RESTAURANT, KENNEBUNK

Having left Boston later than those more ambitious travelers, the bus carrying the Ice Follies fans from Bath made it as far as Kennebunk, midway on the turnpike and about fifteen miles from where Charles Voyer was mired in snow. Stanley Peterson, the driver, was well aware that a Howard Johnson's restaurant sat at that exit, next to an Esso station. Famous for being the first franchise in Maine, it was the only eatery on the toll road and the safest place to spend the night. When the bus pulled into the parking lot, in sight of the restaurant's white cupola and iconic weather vane, a wave of relief washed over the passengers. It was now past one o'clock in the morning and they were tired and hungry. Along the twenty-two miles that Peterson had driven on the turnpike, he counted six wrecked cars and fourteen incapacitated plow trucks. His eyes were bleary from the constant battering of snowflakes against the windshield and the hypnotizing back-and-forth slap of his wipers. He needed to get from behind the wheel of the bus and rest his eyes, maybe even find a place to sleep.

The Bath passengers had a surprise in store when they filed through the restaurant's front door. Beginning as early as nine o'clock that evening, the place had become a refuge for other travelers from the storm. They slept atop their winter coats spread out on the floor. They slept on the wide windowsills, in booths, or sitting up in chairs. Some rested their heads down on tabletops. The door kept opening as more luckless folk came in, stomping snow from their boots. By 2:00 A.M. the restaurant could hold no more people. The Esso station next door was also crammed full. The first to arrive had managed to drive in by themselves and park. Now the parking area was full.

A soldier on leave and traveling home to Wiscasset with his wife and baby daughter pulled into the parking lot hauling a fancy house trailer that was swaying in the wind. It was almost 3 o'clock and he had driven onto "the pike" down at Kittery at 7:00 P.M. Eight hours and twenty-two miles later, he was finally off the road. He had been the envy of the others until he ran out of kerosene to heat the trailer.

The last refugees abandoned their automobiles on roadways and made their way through the heavy drifts, those southbound travelers using the pedestrian tunnel under the interstate. Others were delivered safely to the restaurant by the few police cars that were still mobile, with heavy chains on their tires. Already bone-weary, state and local police officers knew they would be on duty for the rest of the night. Two were at work searching for kerosene for the young soldier.

In all an estimated five hundred men, women, and children were stuffed into booths and crammed side by side along the walls of the restaurant. A three-month-old baby boy who seemed thrilled with the excitement was the center of attention in his mother's arms. The telephone rang constantly, calls from desperate family members having heard a news report about the crowd marooned at Howard Johnson's and asking the whereabouts of their loved ones.

James P. Ivers, the overtired and bewildered manager, kept up morale as best he could. His regular staff of cooks and waitresses had stayed on when the storm intensified, even though most had already worked a full day. Volunteers from the village turned up to offer help in feeding the crowd, some trudging from home through the deep snow.

"Folks, we have enough food in stock," Ivers promised his unusual patrons. "Even if the toll road is blocked for hours, no one will go hungry. The Red Cross knows you're here."

Tabs were kept so that payment for the food could be compensated later. The plan was that meals and snacks would be served only at the counter since the booths and tables were used for sleeping. When one cluster of diners ate, they were immediately replaced with a new one, over again until everyone had eaten. For those who tried to sleep, this constant activity at the counter, the chatter of the staff, and the clatter of plates and utensils made it impossible. Prisons were likely quieter. While the crowd was respectful, even solicitous when stranger met stranger, fatigue was settling in.

As Ivers counted heads, he wasn't sure he would be able to keep his promise. He had already sent a staff member into both bathrooms more than once that night with a plunger. His hope was that the next day would bring rescue vehicles and everyone could get on their way and his restaurant back to normal. With no radio in the place, men slipped out to their parked automobiles to listen to weather reports and then deliver the information to those inside. The storm showed no signs of letting up and was, in fact, intensifying.

In all, over five hundred cars, trucks, and buses carrying a thousand people were now unable to get to their destinations. Churches and farm families up and down the turnpike took in as many needy travelers as they could. For the dozens of less fortunate who were trapped in automobiles not close to towns there was nothing to do but run their engines for a few minutes when the harsh cold seeped in and say their prayers that dawn would bring an army of snowplows to their rescue. It was now after midnight in whiteout conditions. Trying to find a nearby farmhouse in which to take refuge would likely end in disaster. And yet many charitable residents used toboggans and sleds to haul hot coffee, cookies, and donuts to those cars they could reach.

Historic Route 1, running parallel to the turnpike and now receiving its diverted traffic, would soon suffer the same fate. It would be closed down for the first time in thirty years. When all railway travel between Portland and Boston was finally suspended, Maine became a state isolated from the rest of New England.

BRANCH LAKE

Searching for the *Titanic* on a lake three thousand acres big and nearly thirty miles in perimeter, in a whiteout situation on a freezing night, would have been an ambitious undertaking. But Ray Morse and Bob Hogan knew it was their job. Granted, the alternative was sweeter, getting a good night's sleep with their unlaced boots by the front door. Exhausted from a long day, they both knew they needed to be vigilant. Anything can happen out on the expanse of a lake. A couple of summers earlier, a Bangor bank president and his wife had ended up in the water clinging to their upturned rowboat. People had been known to journey onto the ice before it was solid enough to hold them, or their vehicles, and either be rescued or drown. But this was unique: a British Hillman Minx, two decorated veterans of World War II, and a German shepherd dog. It was one for the books.

On snowshoes and leaning into the wind, they spent almost an hour negotiating Hanson's Road leading down to the lake. It was now 3:00 A.M. The wind would be even worse out on the open lake, and carrying with it a stinging snow. Nearly two feet had fallen so far and it wasn't over yet. Not knowing where the men from Brewer were already searching, Morse and Hogan made their way straight out toward Teachers Island, a small piece of land past Hanson's Landing and Camp Jordan. That was where many ice fishermen liked to drop their lines in the water. Maybe the car reported earlier on the Narrows had managed to drive that far north on the lake.

It was almost four o'clock when they saw yellow orbs from flashlights bouncing toward them in the dark. Ray Morse shouted into the wind, asking who was there. Carlton Morrill answered. He and brother Richard, and the two family friends, had gone as far as Teachers Island, thinking Jimmy would have chosen that spot for trout. They saw nothing in those blizzard conditions.

"We won't find them unless we trip over them," said Richard, "not in this storm."

"They're smart enough to stay put," said Carlton. "But we're worried about the temperature. They probably already ran out of gas."

It had been below freezing all day, at 27°F and then down to around 10°F on the lake after midnight. No one said what they were thinking. What if the men had tried to walk out earlier, hoping to make it to the highway before nightfall and were overcome by the storm? With winds at thirty-five to forty miles an hour it was the perfect set of circumstances for frostbite. That meant the wind chill factor would be around 35°F below zero. Frostbite was not uncommon in snow country and a strong wind sped up the process. In extremely cold conditions, circulation is cut off to the most vulnerable areas first, the fingers, toes, nose, and ears. Staying in the automobile out of the wind and staying immobile would be their best decision.

"They'll stay put tonight," said Richard, echoing his brother's words and hoping he was right. Plenty of seasoned men had made the wrong decision in similar circumstances and paid the price. "We'll be back in the morning."

"There were car lights reported early evening down by the Narrows," Ray said. "Maybe they were able to drive a couple miles or more before the snow got them. Bob and I are out here now. We might as well search another hour or two."

"We'll go down the lake a bit," said Hogan. "If we find them, we'll lead them back to the main road if they're up to a walk."

Ray Morse nodded. "There ain't no cars driving off this lake tonight," he said.

Richard Morrill unshouldered a small canvas satchel and handed it to Ray.

"Take this food in case you find 'em," he said. "Otherwise, our wives will kill us."

THE YOUNG SAILOR, HULLS COVE

The deep stillness woke Paul Delaney. He strained to see if anything had changed outside, perhaps the lights of a rescuing plow. He couldn't remember such intense darkness before, not even on those moonless nights after he arrived in Winter Harbor. He had grown to admire the night sky in Maine, so different than the cluttered sky over Staten Island, or in San Diego where he'd been previously stationed. On Schoodic Point the stars seemed to press down and touch the water around Little Moose Island. In these winter months they were so bright and sparkling they looked made of glass that might shatter if touched.

What he *felt* in the car just then, more than *saw*, was an oppressive blackness, as if light had never existed. He had only *imagined* light, in a world that orbited the sun. He sensed the weight of his mortality then, and it frightened him for the first time in his young life.

"Holy mother of God," Paul thought. "How much snow is covering me?"

NEW ENGLAND: A WORLD FILLED WITH SNOW

The outer world, unaffected by the blizzard, went on with life. In Italy a determined group of rescue climbers were scaling the rugged banks of Monte delle Rose, hoping to find survivors from the plane crash that had occurred the day before while anxious soldiers waited in Kenya for news of their wives and children. In Los Angeles, Elizabeth Taylor had a small mountain of luggage packed and ready for her limousine ride to the airport in the morning, on her way to a second marriage in England, one she is certain will be her last. And Willie Sutton, "The Babe Ruth of Bank Robbers," was enjoying one more day of freedom in New York City before two young policemen recognized and arrested him.

But in New England the northeaster had wreaked havoc, with more trouble and sorrow still to come. In coastal cities and towns, electrical wires sagged under the weight of the snow and many snapped, causing power outages. There were traffic accidents, blocked highways, destroyed wharfs, sunk draggers, and battered sailboats. Babies were born in several states that first night of the storm, some of the expectant mothers driven to the hospital in police cruisers. Firetrucks were stymied, snow either preventing them from getting to fires, or hydrants buried deep if they managed to arrive. Dozens of plows went out, hoping to break open the roads but breaking down instead. On that first night of the storm, seventeen people would die in New England, including Harland Davis and Jimmy Haigh. What was to be a peaceful Sunday in towns and coastal villages had become a day filled with grief for some, yet laced with hope for others. Of the two dozen lives still to be lost during the storm, four would be in Maine.

As Sunday night wore on, as the earth spun toward the thin strands of daybreak on Monday, snow was still falling. In the middle of this night, unaware that anyone had died, Bill Dwyer checked his front porch one more time. The snow that had sifted in around the door was unbroken. There was no sign of paw prints. Snooky, his yellow cat, had not come home.

PART THREE

PART THREE

ABOVE: A typical Maine lumber camp, circa late 1940s. BELOW: The American Party, nicknamed the "Know Nothing Party," formed a mob to burn down the Old South Church in Bath, Maine, in 1854. Its parishioners were Irish Catholic and a wave of anti-Catholic, anti-foreigner fury was sweeping the country at the time.

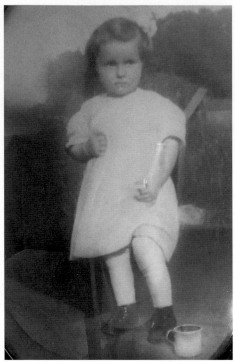

ABOVE LEFT: May Jareny Coombs, who came from an orphanage in San Francisco to Maine as a sixteen-year-old. ABOVE RIGHT: Hazel Coombs, age two, with her blueberry cup on Isle au Haut. BELOW LEFT: Hazel in her uniform as a maid at Point Lookout, the private club on Isle au Haut. BELOW RIGHT: Hazel and Phil, after their engagement. They met when he was ill with pneumonia and she was a student nurse. Fearful he might die, she checked on him often. They married in 1937.

ABOVE LEFT: Harland Davis as Thomaston High class valedictorian, 1941. ABOVE RIGHT: Alice French Church, before their marriage. BELOW: A postcard Harland sent to Alice during their courtship. It shows his wharf at Pleasant Point. The white house at center, with chimney, was the Riley Davis family home. On the back, Harland scrawled, "If you are at the theater, I will know where to find you."

PLEASANT POINT, ME.

ABOVE LEFT: James B. Haigh, a lobster buyer from Portsmouth, NH. ABOVE RIGHT: Alice and Harland Davis, after their marriage in 1950. BELOW: Harland Davis on his boat the *Sea Breeze*, with customers at his wharf in Pleasant Point.

Fabbri Cottage, Eden Street, Bar Harbor, Me.

ABOVE: Buonriposo, the magnificent summer home built in Bar Harbor by Ernesto Fabbri and Edith Shephard, great-granddaughter of Cornelius Vanderbilt. BELOW LEFT: Alessandro Fabbri, a younger brother, was a wireless radio enthusiast. RIGHT: Edith and Ernesto. When Buonriposo burned in 1918, Edith wrote to her mother, "I loved that house and everything in it like another child." It was eventually rebuilt, but then in later years demolished.

ABOVE: The YMCA in Brownville Junction was built when the Canadian Pacific Railroad asked employees to help fund the construction. They would donate a day's pay every year and in return have a place to lodge when working away from home. For locals, it was a center of entertainment, complete with a bowling alley. BELOW: Boy Scout Troop 111, with Ray "Sonny" Pomelow at top left. His friend John Ekholm is at right, in sunglasses. In front, second from the left in the dark shirt is Bobby Williams. Ronnie Knowles is third from right.

ABOVE: Seaman 1st Class Paul V. Delaney at twenty-one years old, and with his only sister, Ellen. BELOW: The U.S. Naval Radio Station at Winter Harbor where Delaney was stationed. Officers occupied the château-esque Apartment Building, designed by John D. Rockefeller's personal architect for his summer home in Maine.

ABOVE LEFT: Peter Godley, in the British 8th Army Signal Corps. RIGHT: James Morrill. The Morrills of Brewer welcomed their English cousin into the family when Pete arrived in Maine after the war. BELOW: From left, brothers Richard, Henry Jr., James, and Carlton Morrill, with the family's beloved dog Laddie.

ABOVE LEFT: Boatswains Mate First Class Weston Gamage, Jr. of the Burnt Island Station. RIGHT: Dr. Charles North, family doctor and Knox County Medical Examiner. North was a physician of the old school, starting his career with horse and buggy. One Christmas Day, a coast guard boat carried him twenty-two miles over ocean swells to tend to patients on tiny Matinicus Island. BELOW: An aerial view of Port Clyde in 1960.

Howard Johnson's Restaurant, Kennebunk, Maine, Midway on the Maine Turnpike

80050

ABOVE: As the northeaster beat away at New England, hundreds of cars carrying an estimated one thousand people were stranded on the 45-mile stretch of new Maine Turnpike. With the first Howard Johnson franchise sitting midway, hundreds found refuge there. BELOW LEFT: A traveler stranded on the pike near the Saco exit. RIGHT: Barbara Ann Haigh of Portsmouth, with cousin David Jackson. Jimmy Haigh's only child, she waited with her mother for news from Maine.

ABOVE: The 1888 blizzard that trapped Mark Twain in a hotel room was legendary for New Yorkers. As part of the clean-up, horses hauled carts of snow to dump in the East River. With elevated trains mired in snow, city planners realized the need for undergound travel. BELOW: When the SS *Portland* went down in a storm that hit New England in 1898, nearly two hundred passengers died. This tragedy prompted ship builders to go from sidewheels and wooden hulls to propellers and steel.

ABOVE: Victor and Myrtle Marden's farm near the Saco exit welcomed in refugees, including Charles E. Voyer, at right. On his way home from Boston after seeing the Ice Follies perform, Voyer became stalled in the snow a dozen miles from the exit that would carry him home. A German shepherd dog belonging to the Mardens stayed busy finding travelers trapped in cars beneath the snow. BELOW: With its chimney built from 100,000 bricks, the Georges River Woolen Mill, in Warren, Maine, employed English-born George Aspey as a carder.

ABOVE LEFT: Ray "Sonny" Pomelow Jr. was a freshman at Brownville Junction High School, an area with a rich history of trains and railroads, sitting not far from the Appalachian Trail's Hundred-Mile Wilderness stretch that ends at Baxter State Park. ABOVE RIGHT: Daniel Speed was the father of a baby and a toddler when he found a steady job in Brownville Junction. BELOW: In Bangor, ambulance men and railroad workers help transport an unconscious Speed to Eastern Maine General Hospital.

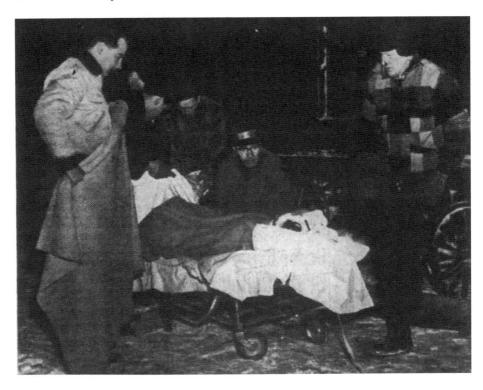

ABOVE: The Mill Syde Lunch served the best pies for miles around. George Aspey and his fellow mill workers often stopped in for lunch or an after-work game of checkers. BELOW: Centre Street in Bath saw thirty-one inches of snow falling overnight, as did many southern and coastal towns in Maine. Plows began breaking down or becoming stuck in drifts as high as fifteen feet.

ABOVE: Dr. Virginia Hamilton, born and raised in Kentucky to a well-established family, delivered hundreds of babies in Bath, Maine. Traveled and well-educated, she was known for her ever-present cigarette. BELOW: Dr. Hamilton, in back in a long coat, and nurse Bernice Brawn, near the toboggan, transport Hazel Tardiff to the hospital with the help of neighbors.

ABOVE: By Wednesday, three days after the first snowfall, the state of Maine began making a full recovery. BELOW: Sixty hours after his ordeal beneath the snow in Bar Harbor, the young sailor Paul Delaney asked to phone his parents on Staten Island. And then he ate his first meal in three days. Weathermen were predicting that another northeaster might be on the way.

FEBRUARY 18, MONDAY

So all night long the storm roared on:
The morning broke without a sun;
In tiny spherule traced with lines
Of Nature's geometric signs,
In starry flake, and pellicle,
All day the hoary meteor fell;
And, when the second morning shone,
We looked upon a world unknown,
On nothing we could call our own.
Around the glistening wonder bent
The blue walls of the firmament,
No clouds above, no earth below,—
A universe of sky and snow!

—"Snow-Bound: A Winter Idyl"
by John Greenleaf Whittier

NORTHEASTERS IN HISTORY

New England has shaped its history on northeasters and hurricanes, a foundation built on grief from the losses and pride in the survivals. Lying beside the North Atlantic, it is bedded down in the very place

where northeasters muster their fuel. With the history of American weather
being a relatively young literature there is no way of knowing when or where
the worst northeaster occurred. But some stand out in weather archives beginning with the "The Great Snow" of 1717. With meagre records to rely on, the
death toll over those nine days in February and March is unknown. What
is known is that this storm dropped five feet of snow on the New England
colonies and the Province of New York. Scores of livestock starved to death
or froze where they stood. More than ninety percent of the deer population
died, either of hunger or as easy prey for ravenous wolves. Apple orchards,
covered in drifts fifteen feet deep, allowed the luckier livestock to graze from
the treetops.

Some reports have houses buried beneath twenty-five feet of drifted snow,
existing only by the corkscrew of smoke wafting up from their chimneys. The
homeowners, unable to get to the woodshed, burned their furniture. Minister
and prolific author Cotton Mather, most remembered for his detestable role
in the Salem witchcraft trials, weighed in on this heavy snowfall of 1717 by
noting that the Boston Puritans canceled church for two weeks in a row. And
Henry David Thoreau, born a hundred years later in 1817, was still thinking
of that legendary storm when he wrote, "The Indians near a hundred years old
affirm that their fathers never told them of anything that equaled it."

The first more carefully recorded blizzard in US history struck with its
greatest force the day after Christmas in 1778, a year after George Washington
and his troops suffered their own killing winter down at Valley Forge. This
northeaster lasted for two bitter days, striking coastal New England hard,
but paying particular attention to the provinces of Massachusetts and Rhode
Island. One of the most heart-wrenching stories in the annals of maritime lore
is that of the *General Arnold*, a privateer captained by the "convivial, noble-
hearted" James Magee, born in County Down, Ireland, and later known as
"the man from Boston." On Christmas Eve, the *General Arnold* and its crew
of 105 men sailed out of Boston Harbor headed for the West Indies. When
they hit a northeaster head-on, Captain Magee sought shelter at Plymouth
and on Christmas Day anchored at the entrance to the bay.

When no boat pilot would brave the hostile waves and guide the *General Arnold* past the dangerous sandbars and shoals, the captain and crew had no choice but to ride out the storm on open water. During the night it intensified, and atop those lethal waves the ship began to drag anchor until she went aground on White Flats. When the tarred seams burst apart after several hours of pounding by the sea, the crew was forced to huddle in subzero temperatures up on deck. There they slowly began to die of exposure. Some downed whiskey after breaking into the storage, hoping it would warm and save them. Magee advised them to pour the alcohol into their boots instead to save their feet, as he had done. Rescuers from Plymouth tried in vain to get boats out to the ship. When they finally succeeded on December 28, only thirty-three sailors were still alive, and nine would perish soon after. A young man named Barnabus Downs, paralyzed by hypothermia and with frostbitten limbs, heard the rescuers refer to him as dead. He managed to blink his eyes before they passed him by.

Of the two dozen survivors from the fated *General Arnold*, a ship named for the man who would soon become a legendary traitor to this country, nine were severely crippled from frostbite. Young Barnabus lost the use of his feet and "walked on his knees for the rest of his life." The deceased sailors were put to rest in a mass grave at Burial Hill, in Plymouth. When Captain Magee died twenty-three years later, in 1801, he asked to be interred in the same grave as his men.

While this blizzard became known in Massachusetts as "The Magee Storm," it was called another name in Rhode Island, hardest hit by its power. The snow that fell there was windblown into drifts as high as fifteen feet. Other boats and ships were damaged but to a lesser fate than the *General Arnold*. Over fifty people were caught in unsheltered places and died of exposure. In memory of the nine German mercenaries who froze to death while standing at their posts in Newport, the storm in Rhode Island became known as "The Hessian Storm."

A child of nature, northeasters are indifferent. This was true of the Blizzard of 1888, which lasted from March 11 to March 14 and trapped the famous writer Mark Twain in his New York City hotel room for several days. Writing home to his beloved wife, Twain complained that he was, *"out of wife, out of children, out of linen, out of cigars, out of every blamed thing in the world that I've any use for . . . a blizzard's the idea; pour down all the snow in stock, turn loose all the winds, bring a whole continent to a stand-still: that is Providence's idea of the correct way to trump a person's trick."*

Meanwhile, an ensnared P. T. Barnum entertained his stranded patrons at the first Madison Square Garden, leased by Barnum and torn down two years later since it was roofless and did not fare well in inclement weather. Both Barnum and Twain were lucky. That northeaster would dump forty to fifty inches of snow from Chesapeake Bay all the way up into Maine, leaving cities like Washington, Philadelphia, Hartford, New York, Boston, and Portland paralyzed in its wake. With winds approaching eighty-five miles an hour, over two hundred boats sank along the Atlantic coast. Thousands of animals froze to death, both wild and on the farm.

More than four hundred people died as a result of that snowfall in 1888. Drifts amounted to forty and fifty feet high, tall enough to bury houses and trains. Half of those casualties were in New York City where newspapers were filled with photos of snowbanks towering over landmarks and storefronts. The East River, running between Queens and Manhattan, froze over enough to tempt travelers to cross it on foot. When the tides changed and the ice broke, many drowned immediately. Others were left clinging to floes where most froze to death before they could be rescued.

Out of tragedy sometimes comes safety and improvement. City planners in metropolitan areas now realized the danger of aboveground water and gas mains, and telegraph lines. They moved them belowground. New York City's elevated transit trains, which had become snarled in snow and were unusable, would a decade later disappear underground for a new subway system, as also happened in Boston. It was a sobering lesson for those energetic big city-dwellers unused to immovability. On the second day of the storm, the *New York Times* wrote, "It is hard to believe in this last quarter of the nineteenth century that for even one day New York could be so completely isolated from the rest of the world, as if Manhattan Island was in the middle of the South Sea."

Such are the bad manners and insensitive behavior of a northeaster. When one hits, it's not a scene set by Currier & Ives.

Some Maine residents now enduring the current northeaster of 1952 were old enough to remember the one in 1888 that had decimated the Atlantic coast and trapped Mark Twain in a swanky hotel room. Many more would remember the storm that hit a decade later in November of 1898, and was closer to home. It was known as "The Portland Gale," named for the steamer it destroyed, the SS *Portland*. The pride of the coastal fleet for the Boston to Portland run, her wooden hull had been built nine years earlier at Bath Iron Works. She was a luxurious paddle steamer with side-mounted paddlewheels, reputed to be safe and dependable on the ocean.

The papers reported that one Maine passenger ready to sail on that fateful night was less impressed with the velvet carpets, chandeliers, and fine china on the *Portland*, and more impressed with the actions of a cat. Before the ship left port, he watched as an anxious mother cat hurriedly carried her kittens one by one down the gangplank and onto the wharf. Figuring the cat might know something the captain didn't, he went inside the ticket office and booked passage for the next day. It played well in the newspapers, but this same prescient cat has carried her kittens off many other ships over maritime history, including the *Titanic*.

Uninformed of the approaching storm, the SS *Portland* sailed from India Wharf, Boston, on the evening of November 27, 1898. She carried upward of two hundred passengers on their way back to Portland, Maine after the Thanksgiving holiday. All passengers died when the ship went down near Cape Ann, not far from Gloucester. While bodies and personal belongings washed ashore for days, the ship itself was not salvaged until many years later when its resting place was finally discovered. Besides the loss of those passengers, another two hundred people would perish in this same blizzard and one hundred and forty vessels would be destroyed. This maritime disaster of the SS *Portland* accelerated the shift from sidewheels to propellers, and from wooden hulls to steel.

ANOTHER NORTHEASTER

By 1952, Mainers remembered many other less famous northeasters as they braced themselves for what was still to come. Snow had fallen all night and was still falling, carried now by gale winds reported in coastal areas of up to 70 miles per hour. And it was not yet over. But folks were used to nature's indifference. They were accustomed to snowstorms. And now, here was the latest one with which to contend. But at least meteorology had advanced beyond a psychic mother cat saving her kittens, and forecasts could now offer the best up-to-date advance warnings. Had Captain James Magee known he would meet a northeaster head-on a few hours out of Boston Harbor, the *General Arnold* would have stayed safe in port that Christmas Eve. And the SS *Portland* would not have left India Wharf loaded with celebratory passengers.

At 5 o'clock that morning, WPOR's 306-foot-high steel transmission tower was toppled by winds and went crashing down into the waters of Portland's Back Cove, putting the station off the air for a few hours until an emergency crew could restore limited power. In those parts of the state hit by the northeaster a similar scene was playing out. Snow had drifted ten to twenty feet high on more than one Main Street in countless towns. It was so deep in the fields along rural highways that residents relied on memory and a telling fence post to help mark the way if a road was even passable. Snow piled high at overpasses, exits, and against bridges on principal thoroughfares. Some farmhouses in the worst-hit areas had it up past their lower windows. Tunnels were dug to outhouses and barns when paths couldn't be shoveled.

With freezing snow snapping electrical wires, hundreds of homes were left in the dark. Telephone lines broke under the weight. Businesses and government offices shut down. Schools were closing and so were factories. No buses or trains were running. Streets were empty except for ambitious shovelers, mostly teenaged boys. New mothers and hospital directors worried how bottles of milk would reach the hundreds of infants and toddlers who would soon be hungry. Eggs were left sitting in crates, the trucks sent to transport them either trapped in snow or unable to drive the roads to make deliveries. Portland, the largest city in the state, had ground to a halt with 1,500 autos left where they became snowbound. Half of the state of Maine was now filled with ghost towns buried in white. And it was still snowing.

JIMMY MORRILL, THE BARTENDER FROM BREWER

It was not yet dawn when a thunderous boom woke Jimmy Morrill. He had been dreaming of the empty shoes again, and that smell of blood mixed with the pungent odor of ammunition. He trembled, waking Laddie, who was asleep at his side in the backseat of the car. Jimmy's first instinct was to grab his rifle. The sound now rolled under the car, beneath the frozen ice. Like a bowling ball, it trundled on to the upper end of the lake. Shaken, Jimmy sat up. It was freezing cold and yet he had perspired, the dream drenching him again in memory. He wiped his forehead on the sleeve of his coat. It was never good to perspire in temperature so frigid he could see his breath.

Pete was asleep in the front seat, although Jimmy didn't know how either of them managed to doze when their fingers and toes felt like pins were sticking them. Knowing the Morrills would come looking for them, they had used the headlights as beacons, sparingly, in order to save the battery. It had proven useless given that snowfall obliterated the beams. At first, they took turns making sure the exhaust pipe hadn't clogged with snow. But around midnight they had run out of gas as Pete predicted they might. The hours following had been spent getting in and out of the car to tread around it in the path they had packed down, shaking their arms to keep the blood circulating, and jumping up and down like two freezing fools.

The sound of thunder on a lake was one Jimmy Morrill knew well when he was not asleep and dreaming of Utah Beach. Even on small lakes like

Branch, ice expansion occurred in winter when the underlying water continues to freeze. As it does, the ice swells and buckles, the sound of which is like the hull of a great ship going aground. When the lake talks in winter, it rumbles and shakes. It's an unnerving experience to a novice. One might think Armageddon was taking place below the surface of the ice. Jimmy remembered Peter's first time to experience this on Moosehead Lake. The Englishman thought the ice was giving way beneath them. The Morrills had teased him about it for days.

Jimmy got out of the car and pushed through shin-deep snow to relieve himself. Before he got back in, he stood a minute and breathed in deeply. The bitter chill wrapped around him. It would be a good time to light a cigarette. But he had given them up on June 6, 1944. That's the day he promised God that if he let him out of Normandy alive he'd never smoke or drink again. He already didn't curse, so that talent was not thrown into the bargain. As a matter of fact, his fellow soldiers chided him about his expressions. "Jiminy Crickets," Morrill had muttered when they hit the first town after coming ashore that day, Sainte-Mère-Église. It had been battered to pieces and what was left standing was on fire. Many of the paratroopers who were sent in ahead of them were still hanging from utility poles and treetops, dead in their harnesses. They had been shot while descending, easy marks when illuminated by the firelight below.

Jiminy Crickets. His fellow soldiers had said little that day, too stunned themselves by the awful sight. It was almost a month after they came ashore at Utah Beach, and after the Battle of Cherbourg. It was at the Battle of Saint-Lô that Morrill and his buddies were in a foxhole with enemy shells exploding around them. One hit so close their ears rattled under their helmets. "Jumping Jehoshaphat that was close!" Jimmy shouted. Hearing this, the soldier they all called Mad Dog turned to him. "You son of a bitch!" he screamed. "If you don't start swearing, I'm gonna throw you out of this goddamn foxhole!"

It was now almost eight years later and he had not broken his promise about cigarettes and booze. And Jiminy Crickets was still his favorite curse. Men should be able to leave a war behind once they've fought it. But the human mind and memory work according to their own devices. Yet, Jimmy knew what his fellow soldiers learned that first day in Normandy. He knew that men aren't born heroic. They do what they must.

The wind sent tumbleweeds of snow rolling down the lake. Then it lay quiet for a minute before it swept in again. Sitting close to his knee the German shepherd whined.

"You hungry, too?" Jimmy asked. Laddie wagged his tail.

Jimmy opened the car door and the dog jumped into the back seat. Cold as it was, the car was still the best shelter. He looked at his watch. It was 6:05, forty minutes to sunrise. He doubted they would see any sun on that day. The snow was still falling. It was already up past the license plate on the car so a little more wouldn't matter. When the motor had sputtered on its last gasoline and died at midnight, Peter had cracked a joke. "It'll be June before a wrecker can come get my car," he said. Jimmy had countered with, "Good, 'cause we'll probably still be sitting in it."

The empty shoes. It was almost this same time of morning in France, 6:30, when he and his fellow soldiers of the 4th Infantry Division had charged the beach in Normandy. Jimmy and the guys knew for days that something big was coming. Stationed in Plymouth, England, while they waited, the military fed them like kings, as though these might be their last meals. They left Plymouth on the USS *Bayfield*, an attack transport. Years later he would learn that a young gunner's mate from St. Louis was on board the *Bayfield* that day, a soldier named Lawrence Berra, later nicknamed "Yogi." Before Jimmy and his fellow soldiers charged the beach, the army had given them a goodbye motto to live by: "If you don't make it, someone in back of you will." The coast guard knew this motto: "You have to go out, you don't have to come back."

The soldiers had crowded onto the landing craft, each carrying a rifle, a bayonet, a hand grenade, and a gas mask they were told might save their lives if needed as a flotation device. The noise was deafening, a constant bombardment of shells. When the wire mesh at the front of the craft went down they hit the water, a wall of bodies. The ocean came up to the waists of some soldiers. Jimmy was five-foot-five-and-a-half inches tall, holding his M-1 above his head. When he thrashed out of the sea and up onto the beach, he fell forward in the sand. That's when he saw the soles of a pair of army boots, all that was left of the soldier ahead of him who had stepped on a land mine. Pieces of body and shards of bone were scattered on the sand, the smell of fresh blood nauseating. "It's for real," Jimmy told himself in those seconds. "There's no wondering anymore. This is what war is like."

Pete woke in the front seat and sat up.

"Top o' the morning to ya," said Jimmy.

"I'm English," said Pete. "But even the Irish don't say that. At least you're still alive back there."

He pushed open the driver's door and got out. Jimmy and Laddie followed. After relieving himself, Pete walked in place near the car to get blood moving in his cramped legs, lifting one foot and then the other.

"It'll be daylight soon," Jimmy said, noticing a thin trace of light on the eastern horizon. "Should we give it a try? Maybe get to Camp Jordan?"

Pete Godley looked up at the swirls of snow coming at him like white sand. Just as quickly, the wind died down. It's what Jimmy had noticed earlier. It was arriving in spurts and bursts from off the highway. When it was daylight, they might be able to see the shoreline well enough between those gusts to make slow headway.

"That means two miles to Camp Jordan in three feet of snow," said Pete. "It won't be easy."

"I got my heart set on bacon and eggs," Jimmy said. He had listened to his stomach growl for most of the night.

"I wonder if the cook can do a full English breakfast," said Pete.

"Let's try it then," Jimmy said.

THE WOOLEN MILL WORKER

Dawn was just breaking. As George Aspey finished breakfast his mother stood at the kitchen window and peered down on the town of Warren. Between gusts of windblown snow, Annie could see yellow lights blinking on in other houses lining the street. Their neighbors were also up early since many of the men were mill workers like George. Gales of wind sifted snow down the embankment. It looked like two feet in the street and covering porch steps. On a clear day, the trees leafless in autumn and winter, the rooftop of the woolen mill with its tall chimney could be seen across the Saint George River. But not on that morning.

"Surely they won't open on a day like this, George," she said. "I can see nothing out there but white." All those years in Maine and yet Annie's English accent was the same as the day the Aspey family had sailed away from Liverpool, fifty years earlier.

"Big companies care more about the color green," George said. He finished his cup of tea. To trek that quarter mile through such deep snow, he and his fellow workers needed to start out before dawn.

George Aspey had expected to rise to a world deep in white but manageable, with the roads being plowed and the sidewalks cleared. As with other Maine towns, Warren knew how to deal with snow. But larger storms were more difficult. The town had no snowplow crew and instead contracted with Chester Wallace to do the plowing each winter. Chester had purchased a couple World War II army trucks that he fitted with plows. But this northeaster put a damper on his efficiency. George had risen to witness out his window a storm that was far from over. He figured Chester was busy plowing out the more important places, like the fire station and the nursing home. Now the workers would be fighting deep snow all the way to the mill. They even had a bridge over the river to cross. But George could find his way blindfolded. He had been doing it for years and was never late for his job in the card room. He just needed extra time that morning to get there.

"Why don't you stop at Ned's house and telephone the mill?" his mother asked. "Might save you making a trip for nothing."

Ned Blackstone, a family friend, had a telephone. He also lived on Riverside Drive, a few houses down. Annie Aspey was worried for good reason. George wasn't a boy anymore, even if he sometimes forgot that when shoveling the porch or putting up their winter wood.

George buttoned his coat and pulled a pair of knitted mittens over his work gloves for extra warmth. Then, smiling, he kissed his mother goodbye.

AT THE SACO EXIT

In the 1947 Dodge sedan, snow up past its tires, Charles Voyer lay covered in the afghan Harriette had crocheted and put in the backseat. His fingers stung inside his leather gloves. He had run the heater as he dared, not taking any chances on carbon monoxide poisoning. He tried playing the radio at intervals for word of what was happening. Were rescue vehicles on the way? But he feared running the battery down and caught mostly static anyway. Despite his heart condition, he was more concerned about Harriette. She would have endured a night filled with worry. A young man in the car stalled behind Voyer's had come a few times to kick snow away from the exhaust pipe, then rap on the window to see if Charles was all right. He and his wife had no blankets or food to offer, but their concern was appreciated.

The countryside was now growing lighter with the approaching dawn. But it was difficult in the blowing snow to discern much of the landscape. Charles almost remembered on his occasional drives to Boston that there were farmhouses near this exit, surrounded by rolling fields. And wasn't there a work farm for indigent folk? Surely a house would be closer than treading through deep snow over two miles to the town of Saco? And then, how much exertion could he stand at his age, and with his heart? Seeing how deep the snow was, he considered the flip side of the coin. How long before he and others would have to wait before being rescued by the snowplows?

Too many years had passed since those days when he was a boy growing up in Manchester, New Hampshire. The world's largest textile mill had been operating there, nearly one-fifth of a mile long. The streets were lined with Victorian houses and the ten-story bank loomed over the town like a modest skyscraper. Charles had enjoyed a happy boyhood there. Running errands at the mill, his dog trailing along after his bicycle, made for good spending money. He was twelve when his parents moved him to Lewiston in Maine, the state of their birth. It was the turn of a shiny new century, and all futures were possible. His father set up office as an optician. When Charles was eighteen, he was hired by the Bates Mill, built at Lewiston Falls to garner waterpower from the Androscoggin River. The looms there were known for having produced textiles made from southern cotton that the Union Army then stockpiled prior to the Civil War. There was a Yankee pride in this that still resonated decades after the North had won and the dirt of southern cotton fields had been blown into the sky by cannonballs.

With his salary from the mill, Voyer was able to take in stage events at the Empire Theater and Central Hall, where it was said that Edwin Booth, brother to Abraham Lincoln's assassin, had once acted on the stage. But it was the impressive Music Hall built in 1877 that he most admired. He found employment there as a flyman at the age of twenty-two, working the curtains and overseeing the scenery. The hall was reputed to be "the best opera house east of Boston." Charles knew early that his passion lay in stage work and the excitement of the theater.

And yet, before he turned thirty, Charlie Voyer was up to his ankles in cold mud in the Argonne Forest in France. There the bloodiest battle ever fought by Americans would play out over forty-seven days of hell. It was the autumn of 1918. In the first three hours of fighting, more ammunition was expended by the Allies than used by the North and the South combined during the entire Civil War. When it was finally over, 350,000 human beings were dead, a good portion from the Spanish Influenza. And Charles was back in Maine, suffering for a couple years from what the doctors called "shell shock."

Voyer opened the console and took out his bottle of nitroglycerin pills. He had been prescribed them a year earlier for the chest pains he occasionally felt. Unscrewing the cap, he flipped one into his hand. He got out of the car, pushing hard the driver's door so that it would open in the piled snow. With his free hand he scooped up a handful to wash down the pill. Harriette had always insisted he never leave the house without bringing at least one in his shirt pocket. The wind still lashing the terrain, he relieved himself quickly and then got back inside the car. He wondered what his wife was thinking then. She would be rising soon for breakfast if she had been able to sleep at all with her husband missing in a blizzard. He had no doubt she would have heard the news and knew that he was somewhere on the Maine Turnpike.

He heard a rapping at his window, the young husband from the car behind. Charles wound the window down enough to hear.

"Good morning, Mr. Voyer." He was tall and dark-haired. He was standing in swirling snow, his eyes squinted against the wind and his neck scarf fluttering. "If some plows don't turn up in an hour or so, my wife and I are walking to a farmhouse."

"I'm thinking the same thing," said Charles. He didn't add that he wasn't feeling well. His watch said almost eight o'clock. He had left Boston fourteen hours earlier. "Otherwise, we'll be here for hours, maybe even a day or two."

"You can come with us," the young man said. "We'll look out for you."

Charles nodded. He knew they were chaperoning him, the old man trapped in the snow ahead of them. But it was as it should be. He remembered his own bravery and that was enough, those days of fog filling the spaces between the trees and pressing down on French farms spread out in the countryside. The damned incessant rain that turned everything into a greasy mud that caked their feet. The cold was intense enough to numb their fingers and toes, but not enough to freeze the mud so that he and his fellow soldiers could finally walk on *terra firma*. They were mostly green, unskilled. Even the officers who commanded them had learned no battlefield tactics to rely on. No place was safe from German artillery, the explosive shells and the mustard gas, not even in their field hospitals or kitchen tents miles from the front lines. They had hurried down toward the fogbanks and the farms that day, shivering in their boots as much from fear as cold, not knowing they were about to take part in the largest battle ever fought by an American army.

"When you're ready," Charles said to the young man who was waiting patiently for a response, "I'll go with you."

HOWARD JOHNSON'S: THE KENNEBUNK EXIT

It looked like a refugee camp. Those travelers who had managed to finally fall asleep by dawn were still sleeping. Many others, too private in their lives to let down their guards in front of strangers in such a public fashion, remained red-eyed and slumped against the walls. Those who had slept washed up in the restrooms where the toilets were constantly being unplugged by the exhausted maintenance staff who had stayed overnight until a fresh crew could arrive. Some of the stranded patrons had bought postcards from the restaurant's rack and devised a game of cards. They dealt them out on the counter, colored images of lobster boats, island sunsets, and lofty lighthouses, as they waited for the first pots of fresh coffee to be brewed.

Acquaintances and introductions had been made during the long night. Strangers were cordial to each other, but there had been minor bickering when lines formed at the bathrooms or the food counters. Had there not been so many pilgrims holed up their stories might have equaled *The Canterbury Tales*. Considering how many of them were returning from the Ice Follies, there was more than one Wife of Bath in attendance. Among the lawyers, mill workers, servicemen, doctors, teachers, nurses, shop owners, college students, teenagers, and a variety of retired folk, was a Maine candlepin bowler returning to Lewiston from a world championship tournament in Boston.

In the busy kitchen the young chef Arthur LeBlanc, who had merely dozed in patches once he had fed everyone, was up and cooking muffins. He and the rest of his weary staff had two hundred already baked with three hundred more being mixed up and ready to pour. LeBlanc was determined to see that no one went hungry. They might wish for a second helping and not get it, but they wouldn't starve. Only twenty-four years old, with a wife and little girl at home and a baby due that summer, LeBlanc was already the star of the show. He and Jimmy Ivers, the manager, were being thanked and praised by their appreciative guests. Whispers were circulating among the travelers that those who could, should put some dollars in a collection plate for LeBlanc and the three cook's helpers who had stayed on after their shifts the day before.

But the snow was still falling outside and the wind playing havoc with it. Occasionally a local trooper pushed in through the front door to give an update on the outer world. The Maine Turnpike and Route 1 were still not open. It was doubtful they would be passable by evening. But at least those five hundred people marooned at the Kennebunk exit, whether at the restaurant or at the service stations on either side of the tunnel, were not sitting in freezing automobiles waiting to be rescued. All they could do was wait for a turn at the counter, savor one of Chef LeBlanc's fresh muffins, and hope the cupola wouldn't blow off the restaurant's roof.

BATH: THE MORNING AFTER

When the city of Bath woke up that Monday morning, it likely wanted to go back to bed. Or it longed for the days when its earliest plows were wooden wedges drawn by sturdy workhorses that tired but didn't break down. As a bleak dawn arrived, two more Chevrolet plow trucks had gone down, one with a blown motor and one a broken axle. Now six of its plows and wreckers were sitting overwhelmed in snow and no sign yet of rescue. This left the city with a pair of light ton-and-a-half plows still working.

The streets and sidewalks were congested with drifted snow, and impassable. A city bus was up past its tires in front of Woolworth's, one of many buses forced out of service. Fire hydrants had disappeared and the tops of parking meters poked up from the snow. Stalled cars and trucks littered the view. Some professionals did their best to carry on regardless. Office owners and workers trudged through snow, some on snowshoes, others on skis. But the business district was mostly deserted. With the city's telephone lines still intact in most places, loyal operators stayed busy.

Highways that led into Bath bringing workers to their jobs at the factories and offices were in worse condition. Seven-foot drifts packed the roads in and around Woolwich, across the Kennebec River to the east. Drivers who had started to work were unable to get as far as the bridge that would carry them across the Kennebec. Once incapacitated by snow, they were also unable to turn around and drive back home. Autos were abandoned along numerous highways. The many smelt huts strung along the Sasanoa River, a tidal channel connecting the Kennebec to the Sheepscot, were now white humps that appeared to be arctic igloos displaced by a tempest.

Hospital personnel in all towns affected by the storm, and knowing the importance of their jobs, did what had to be done to turn up and get things running. Despite the wind and snow, one third of the staff managed to arrive at Bath Memorial. Citizens who lived in the area had faithfully shoveled out the entryway leading to the hospital. Those maintenance crew members who resided close enough to walk showed up at the door, shaking snow from their hats. Cooks and kitchen helpers fought the drifts. Nurses and nurses' aides reported for their day shift. Wearing tall boots and wrapped in heavy coats to guard against the frigid wind, they leaned into the gale, sometimes walking two miles. The night nursing staff, unable to get home, stayed on to assist as the storm raged. At ten o'clock Monday morning they were given hospital beds so they could rest up for that night.

The Bath fire chief made sure he had two men to a truck, ready for emergencies. But it was doctors who garnered the most vocational ink in the newspapers. A police paddy wagon transported one to the hospital at midnight in time to deliver a baby girl. That morning another hiked a mile through snow to perform a scheduled operation. His anesthesiologist hitched a ride in a police wagon that careened into a ditch when a snowplow slammed into it. A wrecker then picked up the doctor and brought him closer to the hospital before it bogged down in snow. The doctor walked the rest of the way, losing his hat in the wind but arriving in time for the operation.

The highway superintendent, Arthur Avery, who had been up all night taking phone calls from desperate drivers about their disabled plows, found his own car stuck in snow on the Winter Street hill. That was the same hill where coasting parties had been held for over a hundred years with Bath children sliding down the frozen slope on hand sleds and barrel staves. Arthur Avery had been one of them, but now he skidded down the hill in his Chevy. Not far from his debilitated car were the city's two sidewalk plows. They had been working overtime to make up for the bigger machines that were decommissioned and were now lodged in snow themselves. "Tell me again why we don't live in Florida?" Avery had asked his wife as he left the house that morning.

VARNEY MILL ROAD

In the predawn hours, Hazel dreamed she was a girl again on Isle au Haut. It was a summer's day with curtains billowing in a breeze through the open windows. She had awakened before her siblings and made her way down to the kitchen where she could smell fresh baking. This had been a ritual in her childhood. Her mother May Jarney Coombs would wake on a weekday morning before the sun had risen over Rich's Cove. By the time the children came down for breakfast, the dining room table would be covered in desserts to last a week. Family meant everything to May, having not known her own mother. She often told her children a story about being in the Catholic orphanage in San Francisco. A pretty woman wearing a fancy hat had come a few times to visit her. "Is she my mother?" May would ask Sister Helen when the lady had left. But she was never given an answer.

When Hazel woke from the dream, she could still smell the fresh cakes and pies. Her mother had been young again, and vibrant. But the island was far away from the lives they now led. She had difficulty sitting up in bed, her lower back sore. She slid her stockinged feet into fleece slippers. It sounded like the storm was still intense. Snow was gusting in from the banks of the Kennebec River and battering the windowpanes. Believing in prayer, Hazel closed her eyes and asked the Lord to watch over her family on Varney Mill Road.

Phillip Tardiff stood at the kitchen window and stared in disbelief at the scene outside. Even though snow had blown off the inclined driveway and down onto the road during the night, the Buick was still up to its license plate. There had to be four feet of snow in the road itself. Where was the plow? He had awakened often in the night to wonder why he didn't hear the noise of it scraping along. Or see the reflection of yellow lights tracing across the bedroom ceiling. Usually, the Bath plows stayed out all night in a bad snowstorm. It was now obvious why he hadn't heard one on Varney Mill Road. It hadn't come.

When Phil lifted the receiver to call Bath Iron Works, he was relieved to hear a dial tone. It didn't matter if they were closing or not. There was no way he could get to work. His supervisor told him they would open, but it was likely they would shut the place down by noon.

"Might as well stay home, Tardiff," he said.

Phil called the highway department next. They were already receiving enough inquiring or irate calls to wear out the telephone lines.

"There are four machines stuck in the snow, Mr. Tardiff," the woman's voice at the town office said. "Two were headed to Varney Mill Road last night and didn't make it." It was not yet eight o'clock and already she sounded fatigued with repeating herself. "I had to use snowshoes to get to the office today," she added, as if her own troubles might soften the complaints coming in.

"My wife is expecting a baby at any time," Phil said, his voice edged with tension. It wasn't this woman's fault. And it certainly wasn't the fault of the guys who drove the equipment. Over the years he had met most of the men who plowed that road, even offered them a cup of hot coffee now and then on a cold day. They had likely been up all night. The only reason they hadn't plowed Varney Mill Road was that nature had stopped them from doing so.

"I'm sure they'll get to you soon," she said, her voice sympathetic.

Hazel came into the kitchen as he was hanging up the phone.

"Did the plow make it out?" she asked. She filled a pan with water to boil eggs, an easy breakfast to make for the children. She felt tired, another restless night.

"It'll be out soon," Phil said. "But I won't be going in to work. Think you can put up with me another day?"

Hazel smiled. She leaned toward him and he put his arms around her, kissed the top of her head.

"Don't you worry about a thing," he said.

The telephone rang. It was Dr. Hamilton. How was Hazel feeling? And had Varney Mill Road been plowed yet?

GETTING OFF BRANCH LAKE

The fishermen cousins had decided to try their luck. They took turns breaking the trail, with Laddie bounding in and out of the tracks they left behind. Between gusts of wind they were able to see shoreline in the distance and they used it for direction. Camp Jordan was still their best destination. If they headed for the highway it meant acres of snow-filled woods to fight through, the ideal place to lose their bearings. The lake was easier, difficult as it was. As they lumbered forward, resting every fifteen minutes, wind whistled through the shoreline trees and spilled down onto the lake, at times sounding like an oncoming freight train. The lake ice rumbled occasionally. On one of their short breaks, Pete lit his pipe and watched the smoke corkscrew into the wind.

"Where did all this snow come from?" Jimmy asked, still amazed that the Hillman had driven easily over the lake just the morning before.

"Snow is still better than sand," Pete said.

"It could be a lot worse," Jimmy nodded. He looked at Laddie. "If we were stranded hundreds of miles in the Arctic with no food, you'd be in big trouble." Laddie wagged his tail.

"If I have to be in a mess like this," Pete said, "it might as well be with you."

Pete Godley felt he owed the Morrills a lot, especially Jimmy, who had provided him with an education in fishing and boating and camping he never would have acquired otherwise. When his five years with the military was up, he gave his mother the pay he had earned, keeping only a few dollars for himself. He figured he could find a job when he got to New Zealand. As a former soldier he arrived in New York City on the *Queen Elizabeth*, free of charge. The plan was to continue on to Auckland, still courtesy of the British government. But when he asked for the connecting ticket, he was told that the

QE offer had expired. Pete was trapped in New York with less than a dollar's worth of coins in his pocket. It was 1948. A policeman found him sleeping on a bench in the passenger terminal at Pier 90 and woke him up. "You can't sleep here, my friend," he said.

That morning, desperate, Pete stopped by a restaurant for a cup of coffee. He saw an ad in a newspaper lying on the counter. *Five dollars and refreshments for every pint of blood.* By the time he got to the address they had run out of refreshments so gave him a shot of whiskey from a flask. He put the five-dollar bill in his pocket. It was a lot of money. He could eat for a time. But a better idea was to find his sister in Towanda, Pennsylvania. She had married a farmer there. They would give him a place to stay and even a job helping his brother-in-law.

When the potato harvest went into full swing, Pete was picking barrels of potatoes alongside the other workers. But they were culling the little ones, opting for the bigger potatoes that would more quickly fill a barrel. Barrels meant money. And then the smallest ones weren't marketable anyway. Pete couldn't do it. He had not forgotten rationing during the war. Or a hunger so sharp in Africa that he had eaten the orange peelings left by his fellow soldiers on the rare occasions they were given fruit. That vitamin C might have saved his life. Others weren't so lucky.

When he realized he would never make it as a potato picker, Peter Godley thought of his aunt and cousins living up near Bangor, Maine. He turned up at their door the day before Christmas and they took him in as one of the boys. In no time he had a home with Aunt Annie Burnett Morrill and Cousin Jimmy. The family also found him a job at the A. J. Tucker Shoe Factory. There was even a girl in the picture now. She worked on the assembly line at the factory and just the week before had approached him. Would he like to go out some time, she wondered? Pete was so shocked he didn't know how to respond. "I'm Margaret Hatch," she said, holding out her hand. "May I call you Maggie?" he asked. "Of course you can't," she replied. "I just told you that my name is Margaret." They went to a movie and a diner afterward for something to eat. He wanted to see her again.

What was noticeable down on the lake was the silence when the wind died down. Fishing there summer or winter they were used to the hum of traffic up on the highway, especially in summers when local kids swam in the lake and tourists rolled down Route 1A. Now there was not even the sound of birds, just stillness unless the wind picked up again and rustled the trees.

"You ready?" Jimmy asked. "Less than a mile and we'll be eating breakfast."

"Knock on wood," said Pete. "If we don't end up in this lake."

"There ain't any wood out here in the middle to knock on," said Jimmy.

Peter pointed. Over the barely visible line of trees on the right shore, an umbrella of snow was billowing up into the air, carried on winds, and headed their way. It reminded Peter of the mushroom blasts made by atomic bombs he had seen in photos. But this was all snow. The cousins stood and watched as it reached its apex and then was carried by the wind down onto the lake where they were standing.

BURIED IN SNOW

The young sailor had no idea how much snow had blown over the top of the car. He prayed that the roof would not be dented or the car damaged in any way. He had promised Jeffrey he would take good care of it. All night he fought to stay awake in case he heard the sound of a plow scraping up on the highway. But what could he do? Push the brake and hope someone saw taillights glowing red beneath the snow? Had he heard the noise of a vehicle nearby, he would have tried anyway. When it was apparent no one would find him that night, he had crawled under blankets in the back seat, and said his prayers.

Awake now, Paul Delaney was back sitting behind the steering wheel. He flicked on his cigarette lighter and held it up as a torch. He tried again to push open the driver's door in case a miracle had occurred while he slept. It didn't budge beyond the original quarter-inch crack. In the light of the tiny flame he saw white snow pressed hard against the window. He opened the side glass and lit a Chesterfield. With his fist he punched a hole in the snow so his cigarette smoke might exit. Smoking calmed him but he was aware it might deplete his supply of oxygen. That, and the fact that the smoke had nowhere to go but back into the car.

Amazingly it still started. But he couldn't run the motor for fear of carbon dioxide filtering in. The exhaust pipe must be packed full. At least with snow blanketing the vehicle as it was, the temperature was bearable. There was no sound of wind. No sound from the outside world. Just that pure, dead silence. He looked at his watch. It was past seven o'clock. That meant morning. It must be daylight. Surely they would find him soon. He had been in the car almost thirteen hours.

PORTSMOUTH, NEW HAMPSHIRE

Bubbles was washing the breakfast dishes. She had convinced Barbara Ann to eat some scrambled eggs and toast. Ellie Haigh could eat nothing. Word had gone out to neighbors and friends the night before. Those few locals who could manage the sidewalk snow dropped in that morning, stomping it from their boots and carrying a casserole wrapped in aluminum foil, or a loaf of fresh bread, or a dozen donuts. Bubbles had arranged the dishes on the kitchen counter, knowing the space would be filled by that afternoon.

Ellie sat at the table staring at the cup of coffee Bubbles put in front of her. It had been a night of dozing from exhaustion, and then dreaming the scenarios of grief. Jimmy had come home and was standing outside their front door in falling snow as Ellie urged him in from the storm. In another, he was calling to her for help, his voice coming from an ocean so gray with fog she couldn't see him. She was now maintaining as best she could for Barbie's sake. She waited until the little girl had taken Skybow outside for his morning walk in the snow before she told Bubbles her decision. She wanted Jimmy to come home one last time. She wanted his wake to be held there on Gates Street, in the house they had recently bought.

"It seems like yesterday we signed the papers," Ellie said. "Jimmy had so many plans."

"Oh, Ellie," Bubbles said. She came to the table and put her arms around her friend. "Waking Jimmy here at home is going to make it harder on you and Barbie. Take some time to think about this."

But Ellie was adamant. She wanted her husband to enter once more through the front door of the home where he had been a loving husband and father when he left.

"We can wake him in the parlor," she said. "Jimmy loved that room."

Bubbles said nothing. It was difficult to argue with a grieving widow. Barbie had come back in from the storm and was taking off her boots at the door. She heard her mother's remark. Her father would be coming home until his funeral?

"Call Wiggin Funeral Home," said Ellie. They were in nearby Dover and had handled the funeral for her stillborn son. "Ask them to take care of details."

Barbie wasn't sure what she felt just then, maybe fear mixed with aversion. A dead body? In their parlor? And then a wall of guilt came down. Her own father. How could she be afraid of this news? She should be happy to have him home again.

"Tell the funeral director I want Jimmy buried with our baby boy," Ellie added.

Barbara Ann moved silently away from the front door and over to the stairs. She ran up to her bedroom, Skybow now following into his newly acquired territory. Barbie closed the door softly so she couldn't hear the voices rising from the kitchen below. This was new information to process. The question that had taken her thoughts away from her father being waked at home was now running wild in her head.

"What baby boy?"

PLEASANT POINT

Alice Davis woke to morning sickness. In the small bathroom of the house trailer she rinsed her mouth with tap water and then washed her face. She felt exhaustion, emotional and physical, as if her body was telling her it was unable to move through another day. Her eyes were puffy and sore to the touch. Before neighbors dropped by that morning, she would put ice cubes inside a face cloth and hold the compress to her eyelids. Still in her bathrobe, she went into the tiny kitchen and filled the percolator with water and coffee.

Like a slow wave, word of the deaths had passed over Pleasant Point, a community that loved and respected Harland Davis. Neighbors had phoned Alice the night before, asking if they should drop by. She told them it was better to wait. She wanted some time alone with Carolyn to process what had happened. She knew they would come that day, carrying pots steaming with chowders, plates of sandwich meats and cheeses, baking sheets of cookies. That's what coastal families had done for generations when one of their own was lost to the sea. The local women would do all they could to help Harland's wife manage her loss. Some had known that loss themselves. Others dreaded it. But Alice felt her grief just then was hers alone.

A cascade of whistling wind and snow swept up the hill from Harland's wharf and drummed against the sides of the trailer. She hoped Carolyn would sleep late. The little girl had listened to her mother cry for most of the night. What Alice had loved most about Harland was how he made her feel safe from the world. Life had been hard, marrying as young as she had, still just a child herself. Carolyn was born two months before her mother turned fourteen. Several months later, Alice lost her father. And now, the new life she had walked into barely a year earlier, her daughter's hand in hers, had died in the ocean with Harland.

It was Riley Davis at the door, that soft knock that always made Alice wonder if it was just the wind. His face told the same story as the night before, that there is no greater loss than this. He stepped inside, standing on the door rug, hat in his hands.

"I come to see if you're doing all right," he said. Alice nodded, tears brimming her eyes.

"It's hard, Riley," she said. "It's hard."

He nodded, his eyes moving to the floor, to the pair of brown slippers that waited there each evening for Harland.

"The funeral place up in Thomaston called," he said. "That's where they took him. They figured I'd be the one to talk to, so as not to upset you. I can go ahead and make the arrangements."

"That's best," Alice said. She couldn't imagine choosing a casket, deciding on flowers, a burial plot.

"I thought we might have a wake at our place," he said then. Alice knew how Harland loved growing up in that large farmhouse with its barn next door and the many animals. "But it would be too much for Eva," he added. He was right. It was hard on them all, and Eva was ill.

"Whatever you decide, Riley."

"They say another storm could be on the way," he added. "Might be best to wait for the funeral service on Thursday. They won't be able to bury him until the spring."

Alice felt relief. It would give them all time to mourn alone, to prepare themselves. Harland would understand.

"I think that's a wise thing," she said. Riley put his hat back on.

"Eva says let us know if you need something."

Alice never doubted that Eva Davis, her mother-in-law, resented her. Harland was Eva's only son, her shiny boy, and he'd already been married and divorced once. Eva was not pleased when he wed a divorcée with a young daughter. It was no secret in Pleasant Point that she never warmed up to Alice being in her son's life. Now that son was gone and Alice was pregnant.

"Tell her thank-you for me," Alice said.

She watched through the window of the trailer door as Riley Davis made his way through the knee-deep snow. He crossed the road that led down to the wharf without looking to see Harland's orange derrick covered in snow. Instead, he walked toward the yellow lights of his farmhouse where Eva was waiting for him. It seemed to Alice as if he carried his entire life on his shoulders.

THE ELLSWORTH GAME WARDENS

Finding no fishermen on the lake, Ray Morse and Robert Hogan had gone home at five o'clock and slept for three hours. Bleary-eyed, they had eaten breakfast and were now back, parked near the sign that said Hanson's Road, its arrow pointing down at Branch Lake. The snow was still falling but the wind was behaving itself. Plows had been able to keep Route 1A open enough for the most necessary of travel. As they were strapping on snowshoes, an Ellsworth police car pulled to the side of the road in front of Ray's car. An officer got out.

"You fellas might as well go on home," he said. "I just got a phone call about them two fishermen."

THE GEORGES RIVER WOOLEN MILL

George Aspey could have done as his mother advised and stopped at Ned Blackstone's house to phone the mill. It would have been wiser than fighting his way through snow for nothing. But even if the mill were closing for the day, George couldn't sit at home twiddling his thumbs. Other workers who lived on Riverside Drive were plodding to the job with him. There was a camaraderie on those walks that he enjoyed, talk of the weather, complaints or compliments on mill policies, or the never-ending subject of sports. But there was little talk that morning with snow and wind stinging their faces as they crossed the bridge.

The mill had opened its doors long enough to inform those faithful workers who turned up that it was closing. The foreman hoped they would be back on course the next day, but that was up to what this northeaster had planned. Some of the men left the mill and crossed the street to the Mill Syde Lunch, George with them. Since the owner lived on the premises the place was open, its yellow lights a welcoming sight. Mill workers often stopped there for a morning coffee. Or for noontime dinner if a lunch wasn't packed and brought from home. At the close of many workdays some of the men enjoyed a game of checkers or cribbage at the Mill Syde as they smoked cigarettes or pipes. The establishment was also known for its excellent pies. George kicked wet snow from his work boots and followed the others inside the cozy cafe.

THE SACO COUNTRYSIDE

Charles Voyer had been thinking of Harriette. She would be worried, maybe staring at the street from their living room window, the way she did when she waited for the mailman. She'd be watching for the maroon Dodge, blinker on and turning into their yard at Cumberland Avenue. He reminded himself that Harriette knew how to handle hard times. Born in a little English village near the Welsh border, she had come to the United States as a toddler. Married at nineteen, she was a widow seven years later with two young children to raise. To support her family, she opened a boarding house.

Charles was forty-eight years old when he rented a room from her. They had been friends at first, taking in a movie together at the theater where he worked, attending a birthday party, or sharing a walk to the corner grocery store for the Sunday paper. They also had loss in common. Eleanor Voyer, his first wife, had also died. Harriette was sixty-one and Charles was fifty-eight when they tied the knot. Who knew they would find love at that age?

He heard the rap again at his window, the same young man.

"My wife and I are leaving now, Mr. Voyer, if you want to come with us."

There was still no sign of a rescue vehicle. Charles took the bottle of pills from the console and slid it into his shirt pocket. He pulled his keys from the ignition.

CAMP JORDAN

They had arrived at Camp Jordan a few minutes before eight o'clock. Smoke curling from the stovepipe on the kitchen's roof had been a welcome banner to the exhausted fishermen. Waiting out the blasts of wind-filled snow had done the trick. Once Morrill and Godley shed their wet coats and kicked off their heavy boots, Jimmy had used the telephone to call the family in Brewer. He caught his brothers about to head back to Branch Lake and take up the search. The Morrills then informed the Brewer police who placed a call to

Ellsworth so that someone would get word to Raymond Morse. *The fishermen are safe at Camp Jordan.*

Route 1A from Brewer to Ellsworth was still being cleared, with a couple of plows breaking down and work coming to a standstill.

"Might as well get yourselves some sleep," Richard told his brother. "We'll come get you this afternoon when the road is open."

In the warm kitchen Frank Lowell, the camp director, was cooking them breakfast. The cousins sat across a rustic wooden table from each other. Pete Godley's first move had been to roll the dial on the camp's big tube radio in search of the world news. He couldn't get the missing Vickers plane off his mind, knowing how far away from England a soldier could feel when stationed in Africa. Those Brits were waiting in Kenya for their wives and children to arrive. Somehow, even wondering when Rommel, the Desert Fox, might pounce next would be less stressful. At least Pete and his comrades were at war and knew the enemy's face. This was to be a joyous family reunion for those soldiers. But there was no mention of the plane.

The cousins ate what the Camp Jordan director put in front of them. The kitchen was usually bustling during the summer months, Frank explained, when the place was open to dozens of happy young campers. The chef and his helper were off for the winter season, but the big fridge held plenty of food. They were served eggs and bacon. Biscuits and jelly. Buttermilk donuts. Plenty of orange juice and coffee. Hot tea for Peter. Jimmy hadn't remembered food tasting that good since Plymouth, England, when the military was feeding them as though they were hungry royalty, a few days away from the invasion at Normandy.

"You getting enough?" he asked Laddie, who had been given a tin plate near the kitchen door with trimmings from cold roast beef and warm scrambled eggs.

"It was a wild night," Pete told the director as he finished eating and lit his pipe. In the cozy kitchen, with the aroma of cooked food and now pipe smoke, it felt like heaven after a night in the Hillman.

"The first cabin next door," said Frank. "I made a fire in the stove and the bedding is clean. Find yourselves a bunk and get some rest."

Laddie followed them out the front door and over to the cabin. Canada jays, aggressive as always, were jockeying for the leftover toast the director had thrown onto the snow. Considered the ghosts of lumberjacks by the old-timers, most Mainers took care to feed them.

"You up for fishing again next Sunday?" Jimmy asked as Pete opened the cabin door. "We can use the Hillman as an ice shack."

RAY "SONNY" POMELOW

When it was announced early that morning there would be no school that day, Sonny had eaten breakfast and then gone back up to his bedroom. He spent some time arranging his Boy Scout badges and awards on the top of his dresser. Troop 111 had held their Court of Honor a year ago at the Brownville elementary schoolhouse next to where he lived. He had been an eighth grader there then. Merit badges and explorer ratings were awarded. Sonny and his friends, Johnny Ekholm and Ronnie Knowles, got the explorer bronze award. But more important, Raymond C. Pomelow Jr. was appointed as assistant patrol leader of the Silver Fox Patrol. It was a special time in Sonny's young life, and he was looking forward to the next Court of Honor in two weeks.

Canceling school for a snow day was rare in Brownville. Missed days had to be made up at the end of the school year when the weather was usually perfect and kids could be swimming at Split Rock, near where the railroad bridge crossed the Pleasant River. Still, Sonny liked the idea of no school, especially on a Monday. But he missed not seeing his friends unless he walked the three-fourths of a mile to Johnny's house. Sonny was always walking somewhere, summer or winter. He had tried saving up for a bicycle. His friends all had one. But it seemed the money he made in the summers picking up discarded pop bottles and taking them back to the store for pennies, or doing chores for the neighbors, was always used for something more important. A pair of new boots. His winter coat. Or if his mother needed help with an overdue bill or the groceries, Sonny helped. He knew how hard Grace worked.

The plow had gone past, but snow was still blowing. There was not much to do in his room except read. He had already studied the pages in the February issue of *Hot Rod*. Fender installation was the cover article that caught his attention. Johnny had been talking about how they might soup up his 1933 Chevy when the weather got warmer and they could work again in his father's shed. That past summer they had fun with the old car. They were lazing around one Saturday afternoon when Johnny found a can of leftover paint in a corner of the shed. Before they knew it, they had painted his car a solid black. Then they opened a can of white and scrawled words like NO GIRLS ALLOWED! on the doors. On the fenders they drew large white eyeballs. There wasn't any artistry to it. They just slapped on the paint.

It was one of the most amusing days in Sonny's young life. He wasn't shy when he was hanging out with his friends. And tinkering with old cars was what the boys did best, where they felt the most comfortable, there and at Boy Scouts. For Sonny it was also an escape from his home life, one that knew drinking and family fights at those times when his mother stood up to Ray Sr. Once, his father was charged with assaulting a police deputy. He pleaded not guilty, but it was determined in court that the assault had been of "a high and aggravated nature." Sonny was younger then, but still embarrassed that it was in the newspapers for everyone to read. His friends never once mentioned it to him. It took two thousand dollars for Grace to bail Ray Sr. out. He later paid a $500 fine, a lot of money for the family.

But there were good times, too, like when his mother and Louise threw Sonny a birthday party. All the invited kids turned up with presents. Louise and Grace even gave him dollar bills inside his birthday cards. Louise was Sonny's guardian angel, not just his big sister. He wished she would move her family back to Brownville. But she and her husband needed to be where they could find work. They now had three small children to support.

Ray "Sonny" Pomelow hated having nothing to do. He closed the magazine and tossed it onto the end of his bed, waking Grace's gray cat that had curled there to sleep. It was not quite eleven o'clock. He could shovel the driveway, but the snow would drift back faster than he shoveled. It was better to wait until the wind stopped blowing. If he lived closer to Brownville Junction, he could go to the YMCA. The Y had everything the local kids needed for entertainment, even a bowling alley. Sonny decided the Y was worth the cold and the snow. If he started walking now, someone was sure to give him a ride. They always did. He might even catch one of the plow trucks. Downstairs, he bundled into his coat and pulled on his boots.

CHARLES VOYER

Snow-filled wind was still blowing, making it impossible to see more than a few yards in the distance. The motorists who had banded together to walk away from their stranded vehicles at the Saco exit chose to go north on Buxton Road. One of them knew there was a chicken and egg farm there, the Marden Place. The young couple who had taken an interest in Charles's welfare the night before stayed close to reassure him.

"They say it's a half mile, Mr. Voyer," the wife said, smiling a lot. They had told him their names, but he couldn't recall what they were.

Charles thought he remembered the work farm for indigents was in the other direction on Buxton Road. It would likely be closer, but he would be safer with a group than on his own. The more robust of the hikers led the way through deep snow and he followed in the jagged trail they left behind. Once off the turnpike exit, the going was more difficult, with small hills ahead. A lot of years had passed since Charles was athletic, even after his military service. He thought of the cold mud in the Argonne Forest and reminded himself that the situation could be worse. "At least artillery shells aren't dropping all around you, Sergeant Voyer," he thought.

But the going was hard and he had already exerted himself during the night, getting in and out of the car to stretch his legs. And then, the soreness in his chest was still there. The young couple finally gripped each an arm, not wanting to embarrass him, he could tell. And yet they saw he was struggling.

"There's the house!" a woman in the group ahead shouted. Her voice seemed an echo to Charles, as if this were all a dream.

"You can make it, Mr. Voyer," said the young man holding his arm. Charles nodded.

"Thank you," he said. It was a struggle to speak.

Ahead was a porch light, a yellow beacon burning in a world of white. A dog came bounding through the snow to the road, barking and wagging its tail. A German shepherd, it looked like. The house was large, a typical New England farm, three buildings that connected to a barn with a cupola on its roof, topped by a weather vane that appeared and disappeared in the gusting winds. Painted white, the farmhouse existed at times only by its green shutters. As they struggled closer, the owner came out and stood on the front steps of a small porch. He held a hand to his eyes to shield them from the stinging snow.

"Come in, folks!" he shouted. "There are others here!"

The German shepherd brushed past Voyer's leg. Charles had had a dog himself once, as a boy in New Hampshire. A medium-sized brown mutt with a spot between his eyes. What was the name? Brownie? He used to chase Charlie's bicycle as he pedaled down the long row of buildings that made up the textile mill. He felt the young couple lift him so that his feet could find the first step. He wished they knew, or that he had told them, "I was once quite a man myself." They were practically carrying him now into a warm room filled with faces. A fire was burning in a woodstove. Charles could smell it vividly, hardwood smoke, a rich scent he loved.

"Put him in that chair," he heard a woman say. The voices were withdrawing, as though he were standing on the back of a train as it left the station. He was aware of faces peering down at him as they fitted him into a chair. He almost smiled to think what Harriette would say, him being treated as if he were a rag doll. He tried then to say *call Harriette*. Why hadn't he told them before, and written down her number? It was too late now. But his wallet held his driver's license. The young wife would know what to do. She would think of Harriette. It felt as though someone had punched him hard in the chest. His breath caught and held. His heart was finally finished and he knew it.

Things would be difficult for Harriette in her old age. She had counted on him. But she was a strong woman.

"Christ, he's having a heart attack!" It sounded like the voice of the young man who had rapped on his car window often during the night. That was a lifetime ago. Faces moved in closer, strangers to him.

"Call Pearl to come!" A male voice.

Charles closed his eyes. He was far away now, looking down on the specks of cars lodged in snow at the turnpike exit, colored confetti placed this way and that. His was the 1947 maroon Dodge sedan. He thought he saw it, its color like dark blood on a white blanket. And then the fields of snow became those rolling hills spread out between the Forest of Argonne and the Meuse River, rising in cold mist, he and the other soldiers with their feet half frozen. "Charlie, I'm telling you the truth," his buddy Vince spoke loudly in his ear. Vince, the big-mouthed kid from Philly. Charlie hadn't seen him since the war. Vince was lighting a cigarette, cupping its burning tip in his hand as he talked. "That trench is filled with concrete bunkers, like a frigging honeycomb. Kraut machine-gun nests everywhere. We'll be ducks on a pond, Charlie." What had happened to Vince? He tried to ask this question. Then Charlie remembered. Vince was still there in France, in the cemetery just outside the village.

"Can you lift him?" a woman's voice was asking now. "Put him on the sofa so he can lie down. Pearl is on her way."

It sounded like the voice of an angel. And that's when Charles saw the six beautiful skaters, the girls from the day before at Boston Garden, floating like seraphs around the rink, the sequins on their gowns shining in the ice like glittering stars.

MEMORIES OF ISLE AU HAUT

Hazel was disappointed that she had forgotten to ask her friend for a perm. Now it seemed that her brown hair, falling just below her ear, had lost its curl. When she brushed it that morning, it had little body. Phil told her it was beautiful so Hazel smiled and accepted the compliment. Who needed styled hair to have a baby? She was alone in the kitchen making a cake, the girls busy upstairs. As expected, Bath Iron Works had shut down at noon. Phil was now splitting kindling in the basement with David helping. The cake Hazel was making was blueberry with a nutmeg sauce, her mother's recipe. Because Phil and the children loved it, it was a goodbye present to enjoy during the time she would be away. Even as a toddler, Hazel would help her older sister pick blueberries in the fields on Isle au Haut. She had her own her tin cup to fill. Still to that day, she could never smell blueberries without memory carrying her back to the island.

She was feeling a loneliness she couldn't explain. She guessed it was the unknown that lay ahead, and that the pattern of the family was soon to change. Hazel knew that change could be both exciting and difficult as new adjustments were made. She also knew that her husband was being protective of her when he said the plow would be along any minute to clear Varney Mill Road. She had called her sister in Bath. "You've never seen such a mess in your life," Evelyn told her. "Plows are broken down all over town and it doesn't look like they'll be back in service for a couple days." Hazel said nothing of this to Phil. But she would telephone Dr. Hamilton later and discuss what should be done. If the baby decided to come that afternoon, or heaven forbid during the night, Hazel was on her own with Phil acting as a terrified midwife.

A song on the radio ended as WPOR began its news broadcast. Despite the storm-battering New England was still receiving, Elizabeth Taylor had landed at LaGuardia and stopped at the gate to chat with reporters, her violet eyes sleepy from the overnight flight from LA. She was on her way to that second marriage in England.

"With everything going on in the world," Hazel thought, "we're hearing about a movie star?" But she wiped her hands on her apron and turned up the sound, pretending it was to inform the girls later since they were huge fans. Liz's flight on to London would leave at four that afternoon. The reporter noted that over a red silk dress the star had thrown on a beige possum coat. Before she was driven to a hotel where she would bathe, wash her hair, and have a short nap, Miss Taylor produced her engagement ring finger for the cameras, a huge sapphire surrounded by two rows of sparkling diamonds.

"Imagine the children we could feed on that one ring," Hazel thought, and switched over to a station in Bangor. The local news was reporting that two fishermen had died off Port Clyde the day before, on their way back from Monhegan Island. One of them was from Pleasant Point, the other from New Hampshire. Her heart ached for the families of those two men. After many generations on Isle au Haut the Coombs family descendants were now all living on the mainland. But they knew from experience that when you make a living from the sea, you might die from the sea. Families the world over knew it. Every weathered port, every wave-lashed dock, every rocky island still carries on its sea winds that eternal voice of grief.

One family tragedy that was passed down through the years concerned Julia Rich Coombs, who was Hazel's paternal grandmother. On a cold December afternoon in 1900, a neighbor ran to Julia's door. Herman and Augustus had met trouble, he told her. "They're bringing 'em in to the dock now!" Herman was Julia's oldest son, just turned twenty. Augustus was her nephew. They had gone out that morning in Herman's dory to haul in their lobster traps. Forgetting her coat, Julia ran following her neighbor over the snowy island road that led to the sea. What did *trouble* mean? She had lost her husband two years earlier when a blood clot traveled to his brain. In an instant she became a widow with seven children to raise. Herman stepped up as man of the house, filling his father's shoes. When Julia reached the wharf, her hair fallen to her shoulders and icy mud clinging to the hem of her long dress, the two boys were lying on the dock. Cousin next to cousin, the life gone from their cold bodies. She knelt beside Herman, dark-haired and dark-eyed, her handsome boy, her firstborn. And dear Augustus, the cousin who admired and looked up to him, thirteen years old that past summer.

In 1926, when Hazel was nine years old, her grandmother Julia died. When the children were old enough to understand the burden of that day, Hazel would pass the family story down to them.

GEORGE ASPEY

Leaving his coworkers laboring over a checkerboard at the Mill Syde, George crossed the bridge. He was obliged to hold his hat in place as wind swept up from between the bridge's columns. Just ahead to his left, sitting on Main Street at the foot of Riverside Drive, was the Masonic Hall. He had held several high ranks in the Masons over the years, including that of Worshipful Master, the highest accorded. He unlocked the door with his key and stepped inside, leaving the storm to lash at the windows. George liked being in the building alone, with its solemn decorum, the walls covered with framed history of the organization's past. At the office desk he read over the minutes of the last meeting. He couldn't telephone his mother, that matter of the Aspeys having no phone. But he called Hilda, his sister who was lodging up on Rod Road as housekeeper for a Warren family. They chatted about the inconvenience of winters in Maine. Hilda was eager for the lively concerts she helped organize in the park each summer.

"June will be here before we know it," George assured her. "And then it will be time to split my wood for winter."

"George, you need to hire a boy," Hilda told him.

"And spend the afternoon teaching him how to wield an ax?" asked George.

He heard Hilda's robust laugh on the other end of the line. He still thought of her as his baby sister even though she would turn fifty that June. Annie was already planning a surprise party for her daughter at the church. Hilda was the only American-born member of the Aspey family, as her Yankee spirit often reminded them.

It was almost dinner time and George considered walking back to the Mill Syde for the beef stew and homemade bread. Then he remembered his mother. Annie would likely hear from neighbors that the mill hadn't opened. Or, if the storm allowed visibility, she would notice from her window that the one-hundred-foot-tall chimney wasn't belching smoke as usual from the boiler room. She wouldn't worry about her son, but her son worried about her. He had even bought a slice of apple pie at the Mill Syde, wrapped in wax paper to carry home to her. Cold wind and snow pushed the door of the hall back into his chest when he opened it. Gusts were rolling down from the hilltop where he lived. George doubted that the mill would open in the morning. He figured a lot of rich men would toss and turn all night, unable to sleep for counting dollars instead of sheep.

He slid the piece of pie into his coat pocket, careful not to damage it. Wrapping his scarf around his neck and pulling his hat down tight, he began the trek up Riverside Drive. With the day overcast and gray, yellow lights burned in the houses on both sides of the street although it was not yet noon. He placed one foot in front of the other, again and again, the drifts above his knees. If not for the snow it would remind him of basic training during his brief time in the army. A naturalized citizen, George was inducted in September of 1918, two months before the war ended. He was stationed at Camp Upton, New York, as a member of the 152nd Depot Brigade. Their job was to organize the new recruits, furnish them with uniforms, equipment, and basic training before they were shipped off to fight on France's front lines.

George had arrived at Camp Upton to find excitement rippling through the ranks thanks to the ruckus caused by a soldier named Irving Berlin. The Russian-born Berlin had already been a famous composer for several years when World War I broke out. So famous was he that newspaper headlines of the day read ARMY TAKES BERLIN! While Irving believed in serving his country, he was also an insomniac not pleased to find himself up at dawn to the sounds of a bugler, and at a salary of $30 a month. So he wrote a song that was soon a hit all over Camp Upton: "Oh, How I Hate to Get Up in the Morning!" Over the years, the song had grown into a national hit. Occasionally, on those early walks to the woolen mill, George would break in with, "*Someday I'm going to murder the bugler / Someday they're going to find him dead.*" He was often joined by his smiling companions who had heard from George about Irving Berlin at Camp Upton.

There was no singing on this day. George could now see his house sitting near the top of the street. Walking to work in the morning was always easier than walking home and he often joked that it should be the reverse, given how tired the men were at the end of their shifts. He wondered if Chester Wallace had one or both of his plows break down that he had not yet opened Riverside Drive. He stopped every few feet to catch his breath, then onward. Finally, just ahead were his front steps. He climbed them one at a time, pushing away snow with his boot. Once he had eaten and enjoyed a cup of hot tea and a smoke, he'd shovel off the porch. He reached into his pocket for the piece of pie.

Annie Aspey was still spry enough to do the cooking for herself and her son. On holidays she made fruit scones and teacakes for the church bazaars. She was sitting in the small parlor, in the chair where she read her Bible. A Christian Scientist, Annie considered prayer as a healer. Having heard the mill was shut down for the day, she knew George would linger in town as he liked to do. The barber shop was near the bridge and Masonic Hall and often the storytelling there on any given day was thick as the hair discarded on the floor.

When she heard his boots on the front porch, she closed her Bible and placed her reading glasses on top. The door opened and George stepped into the hallway. Annie smiled to see the waxed paper he carried in his hand. She often noted that the Mill Syde made the best pies in Warren, even if they couldn't spell. Then she looked at her son's face.

"George, are you feeling all right?" she asked. "You look as if you've seen a ghost."

George said nothing.

"Come in, for heaven's sake, and close the door," Annie said.

The piece of pie wrapped in wax paper dropped from his hand. With the other he let go of the door knob and collapsed onto the hardwood floor.

SONNY POMELOW GOES TO THE Y

As Sonny had hoped, and despite the falling snow, he caught a ride to Brownville Junction in a matter of minutes. A neighbor driving a pickup truck fitted with a plow on the front and chains on the back tires stopped for him. The three-story YMCA was a dream come true for local people of all ages, and in all seasons. Decades earlier, the Canadian Pacific Railway had asked their employees to contribute a day's pay each year toward financing the building. It was finished two years later. It was an idea that would be of service to those employees who stayed there as "roomers" when working away from their home terminals. They had comfortable dormitory rooms on the top floor and in the basement were showers for bathing and a kitchen where they could prepare their meals.

But it was a dream come true for locals. The basement also had a three-lane bowling alley that was available to the public, with local boys resetting the pins. In the main room, there were numerous kinds of entertainment including ping-pong, pool, shuffleboard, cards, and checkers. A snack bar sold candy, peanuts, chips, and soft drinks. Outside was a skating rink. Summer activities were the swings and sandboxes, as well as a basketball and tennis court, a baseball diamond and horseshoe pits. Saturday nights year-round, with the ping-pong table folded and rolled aside, dances were held where kids could jitterbug and waltz.

On that stormy day in February, with several roomers in-house, the Y was still open. Snow days were often manageable since the directors, a married couple, lived on the premises. Brownville Junction kids could trudge through the snow from nearby homes. And Brownville kids would catch a ride for the three miles. Sonny Pomelow spent a lot of time at the Y when he wasn't in school or doing homework and chores. On this Monday, with school closed and nothing but a boring day ahead, he was glad he took a chance on hitchhiking. He pulled open the wide front door and went inside, hoping he might see a kid from his class. Maybe Johnny Ekholm or Bobby Williams had also found a ride to the Y. The building was warm and welcoming. The sounds of bowling pins being struck echoed up the stairs from the basement. A lot of kids like Sonny Pomelow considered the YMCA in Brownville Junction a home away from home.

THE HAIGH FAMILY OF PORTSMOUTH

Ellie Haigh had asked her good friend to contact the Wiggin Funeral Home in Dover. She herself did not have the resilience just then to begin the preparations. Bubbles had telephoned and left what instructions she could. With the storm causing so many disconnects in both states, and with many telephone lines down, information was slow in coming. Bubbles still could not tell the local funeral director just where James Haigh was so that they could make arrangements to bring his body home. She had only a name, and not the town in Maine where the funeral home was located, or even a phone number. She left a message for Tom Barlow at the Portsmouth Police Department to please call her. It had been Tommy who came to inform Ellie of the death. He would know how to get the needed information. But with all the auto accidents, illnesses, and pregnant women needing rides to the hospital, the dispatcher wasn't sure when Officer Barlow would return. But she would see that he called the Haigh residence when he did.

Ellie was now more desperate than ever to get her husband home. She asked Earle Sanders to inform Jimmy's parents of their son's death and he had done so the evening before. They had not yet arrived at Gates Street. There was no need, Ellie told them on the phone that morning, until Jimmy came home. They would only risk their own lives driving ten miles in the storm. She would keep them informed. Mrs. Alice Haigh was not one to break her reserve in public, not even for this. But her daughter-in-law could hear the anguish in her voice. She would send Earle to drive them once Jimmy was home.

With still no information coming, Ellie put her grief aside to get angry. Where was her husband? She could find no place to telephone in Port Clyde. She had no idea who Harland Davis was or even where he lived, only that he was the man Jimmy was to meet. Phone lines were down in many places and busy in others.

"They'll get in touch with you as soon as they can," Bubbles told her, draping a sweater around Ellie's arms. Nothing seemed to get through to her unless Barbara Ann came to hug her. Bubbles saw it happen in a flash as Ellie became the caring mother whose own anguish was now in second place. Her daughter came first.

A dozen friends and neighbors who lived close enough to walk or drive stopped by, bringing boxes and baskets of food. Bubbles put things in the fridge and cupboards, knowing it would be needed if the wake was going to be held there. But still no word of where Jimmy was or when he could be brought back to Portsmouth. Friends filed into the parlor, some sitting on the stair steps, others standing. They all wanted to help in any way they could. Barbie had stayed in her parents' bedroom much of that day, smelling her father's clothes hanging in the closet, his shirts and jackets. And staring at their family picture on the bedstand. Once, she kissed it and whispered to her father that she loved him. Now she had come to the kitchen to say hello to the guests. She was standing near the phone when it rang. After listening to the caller, she cupped her hand over the receiver.

"Mom," she yelled to Ellie, who was now in the parlor talking to Mrs. Mondelli from next door. "It's for you. It's someone in Maine."

What happened next shocked not just Barbara Ann, but the visitors.

"It must be Jimmy!" Ellie shouted. "I knew it was a mistake! He's coming home!"

She rushed to pull the phone from her daughter's hand.

"Jimmy!" she said. "Honey, where are you?"

Barbara Ann turned to the startled faces.

"It's a funeral home," she whispered. "They want to know how to handle the arrangements for Daddy."

MR. VOYER

The Marden home on Buxton Road in Saco was a chicken and egg farm. But with so many refugees there for what might become days Myrtle Marden began calling neighbors as her first weary guests arrived. Thank heavens the phone line hadn't gone down under the weight of snow. Could they donate any food? Her own supplies would soon run out. Roasted chickens and a fresh supply of eggs could go only so far. She figured those folk living within a quarter mile would be able to snowshoe, carrying mostly breads and sandwich meats, coffee, milk, and pastries in knapsacks or bags. But Buxton Road was the only road out to Marden's farm. It would need to be cleared before an ambulance could take away the unfortunate Mr. Voyer, who had died soon after being helped inside the warm house.

When nurse Pearl Grant arrived, she examined Voyer quickly, a matter of procedure for it was obvious to everyone that he had passed on. Pearl found his driver's license inside his wallet. With his name and address now known, Victor Marden picked up the phone to notify the Saco Police Department. They could inform the next of kin. Hearing this, the young couple who had visited Charles during the night to check on his welfare were visibly upset.

"Maybe he should have waited in his car," the young woman said. Her husband put a comforting arm around her. Since one of the other stranded guests was six months pregnant, Pearl turned her attention there before packing up her kit and snowshoeing back home. There was nothing Victor and Myrtle Marden could do but have the body removed from the living room to where it could be kept until the hearse drove the two miles from Dennett-Craig Funeral Home in Saco to retrieve it.

"We'll put him in the back bedroom," Myrtle Marden said, her voice shaking. If she lived to be a hundred she never expected an experience like this in her own home. She went quickly to the linen closet and found two white cotton sheets. There were layers of plastic in storage from when Victor winterized the windows that past autumn. She selected a folded panel and pulled it from its box.

The Marden house was designed as the classic "connected farm" common to New England architecture. The big house had joined to it a little house that served as the kitchen. Next to the little house was the back house, traditionally a carriage room or wagon house that modern families often used as a mud room, or an extra bedroom or storage area. It connected to the barn. Myrtle had turned the back house into a summer bedroom for when family or extra guests might use it on visits. In winters it was unheated and the room stood cold. She carried the sheets there. She lay the plastic panel first on the bed and then spread a cotton sheet on top. The other sheet she left folded for her husband to find and cover the body. Before she returned to the main house, Myrtle stood looking at the room. In one corner were two shopping bags filled with Christmas decorations waiting to go in the attic. Colored blocks of unsewn quilt squares were piled high on the chair. Scattered books decorated the shelf, along with various knickknacks the children had given her over the years. Myrtle Marden whispered a prayer for this man's soul, for the family and friends who would mourn him.

Three men from the stranded vehicles, one the young husband who had befriended Charles, helped Victor Marden carry the body from the living room where Voyer had died, out through the kitchen and into the small back house. With rigor mortis not yet begun, his body was limber, almost lifelike. He was still wearing his winter coat. Dark wet splotches on the legs of his trousers down to his boots was a sign that he had urinated in dying. In one pocket was the stub from his $2 ticket of the day before. *Row F, Seat 6.* They lay him gently on top of the bed. One of the men brought his hat and gloves from the table where Mrs. Marden had laid them minutes earlier.

"I wish we could have done something for him," said the young husband. His voice quavered.

"You did all you could, son," Victor said. He picked up the sheet Myrtle had left lying on the bed and the young man helped him unfold it. They shook it open over Charles Voyer's body and, billowing down, it covered him from head to foot. At the doorway, Victor snapped off the light and closed the door. Charles Voyer lay in the dark as if under a layer of white snow. In the beam of light seeping under the door, Christmas decorations glistened from the shopping bags.

THE MISSING SNOOKY

When Bill Dwyer trudged through snow that morning to his neighbor's front door, he passed bushes and shrubs that now resembled crouching polar bears.

He asked the question he would ask several more times as Monday morning progressed into afternoon and then evening. "Any chance you seen that fool cat of mine?"

Snooky had still not come home. He had been known to rove the neighborhood at times, but that was usually during warm summer nights when plump mice were also on the prowl. Each time a neighbor shook their head and promised to keep an eye out for the yellow cat, Bill would say, "Ah, he's likely just gallivanting. He'll be home when he's past hungry."

Back at his own house, he shoveled the front steps and part of the porch just enough to tire himself. The snow was still coming down so there was no need to put much work into what would soon be undone by the wind. Before he went inside Bill stood staring into the falling flakes. This would be a terrible time for a house fire. Again he felt the loss of his input, that gratification of roaring up to an address to see faces thankful for the arrival of the firetruck. It had given him great pleasure to aid his fellow citizens of Bath.

Bill Dwyer smiled, remembering the only fire he personally had experienced. It was the week before Christmas in 1915. He was still married to Gladys, his first wife, who was young and often naïve. She had dropped an oil lamp and it caught her dress on fire. They lived in a tenement then, over on Middle Street. When the alarm box sounded, the boys from No. 4 had rushed to the scene. Gladys had already doused the flames on her dress, sustaining minor burns. But the couch had blazed up so the firemen unrolled the hose and leveled a heavy stream at it. When Bill got home he found that most of the damage to the sitting room was caused by the water and smoke, not the fire. Gladys cried all that night, with him trying to reassure her. But it set his mind to thinking about becoming a volunteer fireman.

Bill knew personally that life can deal out some surprising cards. The first he had heard after their divorce was that Gladys was remarried and was a clerk in a shoe shop. Then she later wed a younger man, a drummer and actor, and went on the road with him as a showgirl in the Mae Edwards Acting Troupe in Eastern Canada. At least *showgirl* was how the local papers described her. Now, she and her last husband had a taxicab service over in Brunswick. Bill wished her well. There had been a lot of water under the bridge since that night the two of them boarded the midnight train to Boston for their honeymoon. She had been just eighteen and excited as she sat in the seat next to him, the ostrich feathers on her hat moving in the breath of her words each time she spoke.

Bill leaned his shovel near the front door, kicked snow from his boots, and went inside.

RIVERSIDE DRIVE

Annie Aspey stood on her front porch and shouted loud enough that her neighbors finally heard her. Ned Blackstone, the carpenter who lived a few houses away, was clearing his front steps. He threw down his shovel and pushed through the snow to reach the Aspey home. He followed Annie into the parlor to see George slouched in an armchair. The pale blue eyes were open, staring listlessly at the opposite wall. His hat was still on, the signature fuzzy gray hair showing around his ears. Ned knew immediately.

"He fell on the floor," Annie said over Ned's shoulder, her voice frantic. "When he got back up, I thought he would be all right."

Ned unbuttoned the heavy coat and loosened the shirt collar. He checked for a pulse. George's skin was damp to the touch. Helen Blackstone arrived just then and led the flustered Annie into the kitchen. She pulled back a chair so the older woman could sit at the table. Her usual reserve gone, Annie Ashby was crying.

"Please help him," she said. Other neighbors, having heard Annie's shouts, hurried onto the porch and stepped into the hall. Ned looked up and caught his wife's inquiring glance from the kitchen. He shook his head. He turned to one of the men in the doorway. It was Lawrence Wells.

"Larry, go fetch Doc Campbell," he said. "Tell him to bring Cal Simmons."

Annie had kept her crying as personal as she could. But when she heard the name Calvin Simmons, she burst into sobs. Everyone in Warren knew the name of the undertaker.

"And call Hilda," Ned then whispered. "Let her know her brother is gone."

Wells nodded and disappeared out the front door.

"He must be cold," Annie said then and blew her nose. She was struggling to take charge as she always did. She was the one her son relied on for decisions about the house, its maintenance, the electric bill, the grocery list. "He should have a blanket."

Helen Blackstone returned from George's bedroom with a gray military blanket that had been made at the St. Georges Woolen Mill and draped it over his chest.

Wind and snow nearly carried Lawrence Wells down the incline of Riverside Drive. The Simmons Funeral Home that Annie and George could see when sitting out on their summer porch was a right turn on Main Street and up a short hill. Lawrence stopped first at Dr. Fred Campbell's house. The doctor was not at home. He had lumbered through the snow to tend to an ill patient. Wells continued on to the funeral home in search of Cal Simmons. Hearing the unfortunate news, Simmons telephoned Dr. Charles North, the Knox County Medical Examiner down in Rockland, just ten miles away.

The aging Dr. North sounded exhausted when wife Mary passed him the phone. He had been out most of the day dealing with his own patients. But the greater problem was the unplowed roads. They had prevented him from driving down to the funeral home in Thomaston to write medical reports for Harland Davis and James Haigh as he had planned. The newspapers were reporting that the two had drowned when North believed otherwise. Death from exposure was what he intended to release officially. The road over to Warren was also filled in places with a foot or two of drifted snow. The doctor couldn't back his sky-blue DeSoto out of the driveway, much less drive it past the LEAVING ROCKLAND sign. He suggested that Aspey's body be taken to Simmons Funeral Home where Dr. Campbell could examine it later. He himself would drive up and make his report when the roads were open.

After Charles North said goodbye, he pulled the plastic earplug from his ear and turned off his listening device. He hoped this would be the last big storm of his career.

When Calvin Simmons arrived at the Aspey residence almost an hour later, he knelt next to George's chair. Experience told him that Ned Blackstone was right. The eyelids had lost their tautness and the pupils were dilated. George's jaw had dropped, slightly opening his mouth. The first stage of rigor mortis had begun. His body was cold, the blood having drained from the smaller veins in the skin, which had lost its color. Simmons knew it as *the death chill*. He went into the kitchen to pay his respects to Annie Aspey, who had always been active in community functions, as had Hilda and George. It was the family's way of showing their gratitude to the new town that had welcomed them from England so many years earlier.

"He was a good man, Annie," Simmons said. "A hard worker."

Back in the parlor, he nodded at Ned Blackstone, who had decided to stay until George could be taken away. He owed Miss Aspey that, he and Helen. They sometimes joined the Aspeys, with Hilda, too, for talks on the porch when the summer weather was pleasant, and a breeze wafted up from the Saint George River.

"It might be tomorrow before we can drive our hearse up this hill," said Cal. "I sent word to Chester Wallace that we need one of his plow trucks. I'll be back in less than an hour to get the body."

Ned Blackstone said nothing.

LEAVING CAMP JORDAN

It was past noon when Jimmy Morrill woke inside the cabin. He kicked a foot up at the bunk above his to wake Pete Godley.

"Hey, sleeping beauty," Jimmy said. "It's time for us to go home."

Pete sat up and swung his legs over the side of the bunk.

"I haven't sleep that well in a long time," he said.

In truth, Pete had dreamed of Tobruk again, the awful heat and sand. General Rommel was always on their minds back then. He was aggressive, headstrong, but you had to respect the son of a bitch and the British soldiers did, even if their leaders didn't. In return, Rommel respected them and had said so. He once told a defeated British unit, their pores filled with dust and their eyes red-rimmed with sand, to hold their heads high. They were valiant fighters he would be honored to command. He asked if they had complaints of how German soldiers were fighting. He was the kind of military leader Pete wished was on the side of the Allies, and not the atrocities. He was still with the British 8th when they heard the news, in October of 1944. The Desert Fox had been found guilty of conspiring to assassinate Hitler. He had been given a choice: die by suicide and receive a hero's funeral or be publicly executed and dishonored like Mussolini. No one doubted what choice he'd take. The world heard the details later, how the car sent by Hitler drove up to the Rommel home as he said goodbye to his wife and son. A few miles outside the village, the car stopped so that he could take the cyanide pill he was given.

"Let's make ourselves pretty," said Jimmy. "We're going home."

The fire they'd made in the stove that morning had gone out. They poured cold water from the jug sitting next to the basin and shivered as they took turns washing their faces.

"First thing I'm going to do is shave," said Pete. For those many months he had spent in the desert, he and his fellow soldiers were given a quart of water a day for cooking, washing their clothes, and shaving. They grew beards until the Sergeant Major gave an order. "Corporal Godley, get rid of it." Pete and the others saved two inches of weak tea in their mugs and shaved in that. Since it was in greater supply than water, they dipped their clothing in gasoline to kill the lice. They didn't realize that the difficult times wouldn't begin until Rommel arrived at Tobruk with his tanks.

Jimmy opened the cabin door and let Laddie bound outside to relieve himself in the snow. A light was burning in the kitchen cabin and they stomped toward it over the path they had made that morning. It had filled in again while they slept, but the wind had shifted and was no longer blowing from off the lake. Snow fell lightly compared to that morning. Again, Canada jays and chickadees fought for the stale biscuits that had been tossed to them. A red squirrel raced away from the food and up to the branch of a nearby birch, barking a warning, its tail jerking. Without the incessant wind, the sounds of nature were returning.

Frank Lowell, the director, smiled when his two unexpected guests kicked snow from their boots at the door.

"If the highway is open my brothers will come get us," Jimmy said.

Frank motioned for them to pull back chairs at the table. He put a plate of ham and cheese sandwiches before them, the homemade bread thick and soft. There was a pot of hot water on the stove, and he tossed in tea leaves for Pete's tea. Another pot held strong coffee. He poured Jimmy a cup.

"We want to pay you for everything," Jimmy said, and watched as Frank shook his head.

"You'd do the same for me," he told them. He put a bowl of pickled eggs onto the table and paper napkins. A plate of sugar cookies was next and then he poured a cup of coffee for himself. "It's good to have company," Frank added. "It's pretty quiet out here in the winters until the camp opens this spring."

Jimmy telephoned Cap Morrill's and his brother Richard answered the phone.

"We're just getting ready to drive over there," Richard said. "The road ain't pretty, but the plows are running. We're taking your car."

"Sounds good," said Jimmy, smiling. He knew his brother was joking. His car was a 1950 Nash Rambler, his pride and joy, a two-door compact sedan not known for its prowess in the snow, or a comfortable ride for four men and a German shepherd. They'd likely be coming in Carlton's big Chrysler.

"Be careful on the turns," Jimmy said. "That little Rambler can be tricky."

Frank Lowell had been listening.

"Our road down to the camp won't be plowed for some time," he said. "But I can loan you each a pair of snowshoes."

When Frank went to fetch them, Jimmy took out his wallet and slipped a five-dollar bill under his plate. It would more than pay for the food and even lodging. He had already told Frank to stop by Cap Morrill's some Saturday night, that everything would be on the house.

When the two men finished eating, they strapped the Camp Jordan snowshoes onto their boots. Laddie would have to follow in the tracks they left. Saying goodbye to Frank, they began the trek along the edge of the lake to Hanson's Road and the highway, a mile and a half away. It was just before two o'clock when they reached Route 1A to find Carlton and Richard Morrill idling at the roadside in Carlton's monster Chrysler.

Laddie sat between Jimmy and Pete in the backseat as the car spun its way onto the road and headed toward home. Snow pelted the windshield.

"I never saw anything like it before," said Jimmy.

"That snow came out of nowhere," said Pete. "One minute we were fishing and the next we couldn't see shore."

"You had us worried all right," said Carlton.

"Thanks for looking for us," Jimmy said. "Must have been a long night for you guys, too."

"It's what brothers do," said Richard.

As the warm car covered the twenty-five miles back to Brewer, snow slapping the fenders, Pete Godley said little else. He was thinking of Richard's words. *It's what brothers do.* He had gone looking for his own brother once. It was morning at Tobruk that he and his fellow soldiers watched as plumes of dust, orange and pink, rolled across the horizon like distant feathers. They knew by then what a dust storm looked like and it was no dust storm. Then the radio transmission came. *A whole bloody German armored division!*

That meant panzers. Peter was in the Royal Corps of Signals, assigned by the British 8th Army to the Coldstream Guards as combat support arms for communication. The Guards had been instructed to surrender at Tobruk. They didn't. With Rommel on the horizon they beat a retreat across the scorching sands. Peter knew he was leaving behind his older brother, Jim, who was also in the Signal Corps, but assigned to another division. Once they had safely rejoined the 8th Army, he asked for a Jeep to head back in the direction of Tobruk. He had to see if he could find Jim. That night he pulled up next to a

Jeep broken down on the roadside, two soldiers lying asleep inside. For years it had been their joke. In the moonlight Pete had recognized his brother's white legs, covered with desert sores and propped up in the vehicle as he slept. Jim had also escaped Tobruk.

"We're home, boy," Pete heard Jimmy say. Laddie began an excited dance to get out of the car. They were just pulling into a parking space at Cap Morrill's tavern.

THE ASPEYS OF WARREN

It was still early afternoon when Chester Wallace maneuvered his plow truck up Riverside Drive toward the residence at the top of the hill. Hilda Aspey was in the passenger seat. When Chester drove over to Rod Hill to pick her up she had been crying. Now she seemed resigned to keep her composure. She would need to comfort her mother. Annie had doted on George, her youngest son. Chester watched as Hilda jumped down from the truck and waded through drifts to the front porch. Helen Blackstone was reaching out a hand to help her up the steps. Calvin Simmons had returned, again by walking up the hill. He had brought with him his assistant at the funeral home. The two men and Ned Blackstone were waiting on the front porch. Cal made a signal to Chester to turn around and back the truck up to the steps. Chester ground the gears as he put the deuce-and-a-half in reverse.

Known as "the workhorse of the army," these trucks were praised by General Eisenhower and other officers for their help in winning the war. That's why Chester Wallace purchased two for his construction company, and his plowing contract with the town of Warren in the winters. Used to carrying everything from bulk food supplies to ammunition, the truck would now bear the small body of George Aspey down the hill and over to the Simmons Funeral Home.

Neighbors in nearby houses watched politely from behind window curtains as George's body was carried out onto the porch. Without the cadaver bag that would have arrived with the Simmons hearse, Cal had instead wrapped George in blankets from the woolen mill. Chester got out from under the wheel and helped the others carry the body down the snowy steps and slide it into the canvased back of the truck. Cal and his assistant climbed inside to stay with the body, out of respect to George and the Aspey family. Back in the cab, Chester turned the ignition and the historic truck roared to life. Curtains fluttered on each side of the street as the plow truck rolled down toward town, leaving snowy tire tracks in its wake.

MONDAY AFTERNOON

Unwarmed by any sunset light
The gray day darkened into night,
A night made hoary with the swarm
And whirl-dance of the blinding storm,
As zigzag, wavering to and fro
Crossed and recrossed the wingèd snow . . .

—"Snow-Bound: A Winter Idyl"
by John Greenleaf Whittier

TRAINS, PLOWS, AND WHISTLES

The steam locomotives in Maine stayed busy in all seasons transporting passengers to their destinations, or carrying lumber, potatoes, and other products to market. An important aspect of railroading was keeping the tracks cleared and in good working order. Winters could pose a problem for trains in those American states and Canadian provinces known for heavy snowfall. A storm could interrupt and even shut down train schedules. Because railroad companies had their own plows, they had the upper hand on travel during harsh winters. Powerful, the plows cleared towering drifts away from tracks and embankments.

A plow train looked like a caboose car with a built-in wedge for a face. Since they were not motorized but pushed instead by the engine, these plows were operated by a plow master, not the engineer. As a matter of fact, the engineer saw nothing ahead on the track when a plow was engaged, even though he ran the locomotive. The plow master sat at the top of the car in a small cupola that had sliding glass windows. From there he watched the track ahead and operated the air valves for the whistle, the emergency brake, and the plow wings. Being a plow master was a demanding job during the worst storms. Because the men were often gone for days each winter, there were small living quarters inside the car consisting of a bunk bed and an oil stove. The plow master also cleaned snow from around the whistle when its mechanisms became clogged. It was his job to blow the whistle at crossings.

People living near railroad tracks, in towns or cities, grew up to the music of these belching locomotives and their steam whistles. Many enjoyed the lonesome refrains, finding in them a nostalgia for bygone days just as they were being replaced by diesels. The sound of a train whistle can carry for miles around. It *needs* to travel since the enormous weight and momentum of an average freight train running at fifty-five miles an hour will require a mile or more before it can stop. And yet that same whistle might not be heard inside an automobile with windows wound shut, with the engine running, with the radio playing, with conversation going on among passengers in the car. It was often noted by residents who lived hundreds of yards from a railroad track that the whistle was better heard inside their kitchens or bedrooms than inside the cab of an automobile idling at the crossing.

In small towns a grade crossing—where a railroad track crosses a road or highway—demanded attention, especially if the trains did not run on a strict schedule. Not hearing the warning whistle had swift consequences if a driver crossed in front of an oncoming train. In 1952, the grade crossings at Brownville and Brownville Junction did not have safety gates or visual signals, those "traffic lights of the railways." Instead, it was the engineer's job to blow the warning whistle at each crossing. If a plow train was engaged in the wintertime, the plow master blew the whistle. In railroad operating manuals it's known as *Rule 14L*, a pattern of two long blasts, one short, and one long.

DANIEL WESLEY SPEED

Danny Speed was twenty-four years old and already raising a family. He was born and educated in Bradford, twenty miles from Brownville. He left during his senior year of high school to enter the Merchant Marine. Like many of his friends, he was a young man eager to serve his country in wartime. In May of 1945, the day he was to leave home, he was instead drafted by the army. The war was pretty much over by then, anyway. When he was discharged in 1946, he came home to Bradford to help his father manage the farm. Two years later, he met a beautiful girl who lived in a nearby town. Danny knew from the start that he would marry Frances Harris. Six weeks after their first date, on a cold December day in 1948, they said their vows. He and Frances set about raising a family in his hometown of Bradford.

By February of 1952, Daniel Speed and his wife were the parents of two boys, one a toddler and the other just a few months old. Realizing he could not support a growing family on a farmhand's salary, he went looking for a better situation. He found it with the Bangor and Aroostook Railroad as a section man whose job was overseeing a section of track by replacing broken rails, tightening bolts, and tamping in new ties in place of rotted ones. If he worked hard, he could move up the ranks. He had his sights set on being a railroad fireman or "oiler," responsible for the train's fuel, oil, and other maintenance duties.

Traveling to work in Brownville would mean a round-trip of fifty miles a day, not a promising drive on a narrow highway and especially in the winters. Daniel and Frances hunted for a rental closer to the job and found an upstairs apartment in Brownville. It was in the house of Bill and May Thomas, who happened to be Sonny Pomelow's maternal grandparents. It was a good enough job with the railroad that the Speeds bought a newer car, a 1947 Ford. Frances kept the small apartment clean and cozy, and the young couple settled down to a better future ahead.

THE HITCHHIKER

That Monday afternoon Danny Speed was pulling out of the parking lot at Bryant's drugstore in Brownville Junction. He had gotten off work in Brownville at three o'clock and driven to the store to pick up medicine for the cold he felt coming on. He didn't look forward to battling a cough while on the job, especially in winter. Even though the plow had cleared the street earlier, new drifts thumped against the oil pan of the Ford as he pulled away from the curb. Behind him, the orange and blue Rexall sign over Bryant's door swung back and forth on its metal pole. The wind was picking up. With snow still falling, the day was overcast and gray. Stores and houses in Brownville Junction had lights burning early in the windows.

Danny was just leaving town when he saw a boy walking along the roadside. Between gusts of snow he recognized Sonny Pomelow's green and black winter coat and his gray woolen cap. Danny felt sorry for the kid. Ray Jr. often seemed lost, maybe lonely, and was usually in need of a ride. As he shifted into second gear to pull over, Sonny turned toward the vehicle and stuck out his hitchhiking thumb. He smiled when he saw it was Danny Speed stopping. In seconds, he had opened the passenger door and jumped inside.

"Sonny, why are you walking in this storm?" Danny asked.

Sonny was kicking snow from his boots. He held his mittened hands in front of the heater vents.

"It's sure coming down, ain't it, Danny?"

"A car could hit you," said Speed. "It's hard to see in all this snow." He liked the boy, who was quiet and polite, even if he wasn't the smartest. He was a good kid.

"No school today," said Sonny. "So I got a ride to the Y."

Speed pulled the Ford back onto the road, his windshield wipers fighting snow. Less than three miles and they'd be in Brownville. He would drop Sonny off at the Pomelow house a half mile farther up the road from his rental, unless the boy wanted to visit his grandparents. Danny was anxious to get home before dark came on and the storm got worse. Frances had a pot roast planned for supper. He could spend time with the babies while he listened to her bustling about the kitchen, pleasant aromas filling the apartment. Once he had shifted back into third, he reached for the pack of smokes he kept in his shirt pocket. He lit a Lucky Strike and threw the dead match into the ashtray. Smoke curled from the tip of his cigarette. Sonny watched him, curious.

"L.S.M.F.T.," said Sonny. "Lucky Strike means fine tobacco."

Speed smiled. Sonny was often saying things like that, ideas that made you wonder why he knew them or where he learned them. But the cigarette ads were everywhere, especially as commercials for the most popular television shows with big Hollywood stars approving the brand as they lit one up.

"I hope you haven't started smoking," said Speed. He drew on his cigarette. Drifts of snow occasionally smacked the bottom of the car as it rolled ahead. The plows would have a busy night.

"Naw, I ain't smoking," said Sonny. "They got a big colored poster up at the drug store. A cheerleader kicking her legs and waving pompoms. L.S.M.F.T. She's kinda pretty."

"That's good," said Danny. "You got plenty of time to smoke if you decide to."

"The Boy Scouts don't want us to smoke," Sonny added. "That's okay because me and Johnny Ekholm found a pouch of tobacco last summer and tried some of it. It made us sick to our stomachs."

Danny Speed smiled again as he kept an eye on the road. At times when the wind rolled in, carrying layers of snow, it was difficult to see beyond the car's hood. The wipers worked hard and fast keeping the windshield cleared. He stubbed his cigarette in the ashtray and held the steering wheel with both hands. Ahead he saw red taillights. They were welcome beacons in the storm and he slowed to stay behind them. The vehicle was a 1949 Dodge panel truck. Daniel Speed followed it over a grade crossing as they entered Brownville.

"Looks like Albert St. Louis got caught delivering mail in this storm," he said. A few hundred yards ahead were two more grade crossings. Danny always lectured Frances to be extra careful at crossings when she drove to Brownville Junction to shop for groceries at the A&P. She knew to look both ways.

"Albert's got a peg leg," said Sonny. He grinned at the notion. "Dad says that's why he's got a permit to hunt from the window of his pickup truck."

"That so?" asked Danny. He had met St. Louis a few times and noticed the narrow wooden leg he thought only pirates wore.

"Dad says Albert should be hunting woodpeckers instead of deer," Sonny said, and was pleased when it made Danny laugh.

"He won't meet any woodpeckers in this storm," said Danny. "But I wish him luck if he's on his way back to Bangor."

Albert St. Louis picked up the mail bags in Bangor six days a week and delivered mail to Milo, Brownville, and Brownville Junction.

"Neither snow nor rain nor heat nor gloom of night," said Sonny.

"Where'd you learn that?" Danny asked. "That's pretty smart stuff, Sonny."

"Boy Scouts," said Sonny.

It was 3:45 P.M. when Albert St. Louis slowed for the next grade crossing. In the Ford's headlights, snowflakes twirled like white moths. Danny and Sonny watched as the mesmerizing red taillights drove onto the tracks and off the other side. In the panel truck's wake a cloud of snow-filled wind blew in over the rails. Danny looked both ways and saw nothing but white up and down the track. He gave the Ford some gas and followed the mailman.

A hundred feet away and chugging north on the tracks, its black smoke corkscrewing into cold air, was a B&A plow train. It was an extra, and not on a scheduled run. If either Daniel Speed or Sonny Pomelow heard its steam whistle, they heard it too late. But the plow master on Engine No. 46 had executed the two long, one short, and one long blasts as the train approached that Brownville crossing.

It had been a day of snow and wind, with no visibility to count on, with sound existing on another planet. The train was plowing the track north to Millinocket. Slowing to twenty-five miles an hour as it approached the village, it struck the driver's side of Daniel Speed's car just behind the headlights. In a matter of seconds, it pushed the vehicle a hundred and fifty feet along the track, slamming its passenger side against a telephone pole, which snapped fifteen feet from the ground. The Ford rolled another seventy-five feet until it landed on its roof and skidded to a stop down the embankment.

The harsh sound of crumpling metal echoed back from the frozen trees beyond the track. Because of blowing snow, the plow master was unable to see the crossing from his cupola at the top of the plow. Realizing instantly from the jarring impact that the train had collided with a vehicle, he pulled the emergency brake valve. The wheels grated steel on steel along the tracks, sending up orange sparks and tufts of snow for a half mile until the train was able to stop.

EMERGENCY CALL

William Sawtell Sr., who lived near the first grade crossing coming into Brownville, was out shoveling his yard while his two small sons watched from a window. Bill recognized the whistle of the approaching B&A train a quarter mile away as it slowed for the grading. Then came the grinding sound of metal crunching against metal that resonated over to the small hill where the Sawtells lived. He knew immediately what had happened. He dropped his shovel in the snow and ran up the steps and into his house.

"Connie, call the operator!" he shouted. "The train just hit a vehicle on the second grade!" He had no idea if it was a car or a truck, but being a B&A train man himself, he knew it was bad. Brownville had no ambulance service so Connie quickly rang up Alice Graves, the telephone operator, who would in turn telephone Charles Foulkes, the local deputy sheriff.

"I need to get down there," Bill Sawtell told his wife, and disappeared out the front door.

Other residents came out to their front yards when the awful sounds reverberated inside their houses. They knew as well what it meant. They could only wait for word, or wade through deep snow to the crossing and see for themselves. Some neighbors did that, men wanting to offer what help they could.

Alice Graves caught Charlie Foulkes just as he was sitting down to a cup of hot coffee and a donut before heading back into the storm.

"I'm on my way," Charlie told her. Before he could get his coat back on and buttoned up the phone rang again. Alice had just gotten a second call, this time from the Melanson family who lived nearer the tracks where the wreck had occurred. The train had hit a car and there were people inside. Charlie told Alice to phone the hospital down in Milo, four miles away, for an ambulance and a doctor. The hospital then contacted the Keniston Funeral Home to send their Cadillac that served as both ambulance and hearse. Dr. Carde, who was on duty, would be waiting for it at the front door.

Albert M. Carde was another graduate of the Medical School of Maine at Bowdoin College, where Sarah Orne Jewett's father had been a mentor to many young medical men in his day. It was also where Dr. Charles North had studied medicine. Carde received his degree from the school in 1919. He had moved to Milo in 1925 and settled down there to practice medicine. Minutes after the hospital received the emergency call, the doctor and his medical bag were waiting for the ambulance. Keniston's had engaged a plow truck to lead the driver of the ambulance over the heavily drifted road.

As the vehicles sped north to Brownville, Dr. Carde lit up a Camel and watched snowflakes fluttering into the headlights. The windshield wipers worked furiously. Train accidents around the country killed or injured a few thousand people a year. There was always some zealous driver trying to beat a train across the tracks, or one who was preoccupied and not paying attention. There had been ten deaths and fourteen injuries in Maine alone in just the past several months. Albert Carde wasn't sure what he'd find at the Brownville grade crossing.

SCENE OF THE WRECK

Parking his car near the crossing, Deputy Sheriff Charles Foulkes hurried down the track and then waded through snow to where the Ford lay. The wreck was almost a half hour old and a small crowd had gathered on the tracks above.

"That's Danny Speed's car," someone shouted. "He and his wife live upstairs at Bill Thomas's house."

With winter's early darkness coming on, Foulkes leaned down to shine his flashlight into the driver's window. The Ford was so badly smashed he figured it would need to be turned back onto its tires before any rescue attempt could be made. It looked like Daniel Speed was still alive. But there was so much blood on the man's face, Foulkes wasn't sure. He reached through the driver's broken window and touched Speed's hand. When he heard him groan in pain, he knew he was right. Shining the flashlight over to the passenger side, he recognized right away who it was, despite the blood and the damage done to his body. It was the Pomelow kid who was often looking for a ride to and from Brownville Junction. Thinking he would be sick to his stomach, Charlie stood and leaned back against the car. Sonny was a good kid. Everyone liked him.

"Get me some blankets!" Charlie shouted up to the spectators. He needed to cover Speed. The day was growing colder and he had no idea how long it would be before they could get him out of that car. He figured they'd care for Ray Pomelow Jr. last. Charlie was no doctor, but he couldn't see how the kid was still alive.

Long minutes passed before they heard the wailing siren of the Milo ambulance and saw its twirling red light. It pulled in as close to the tracks as the driver could manage in the snow. Two of Brownville's volunteer constables also arrived.

"Dr. Carde is here!" someone shouted, and Charlie felt relief wash over him.

The tops of his galoshes filling with snow, Dr. Carde followed the constables over the tracks and down to where the car lay on its roof.

"The driver is still alive," Charlie told him. "At least he was a few minutes ago. His passenger is a kid named Sonny Pomelow."

Charlie shined the light into the car as Carde knelt and reached for Danny Speed's wrist.

"He's alive," he said.

Charlie then directed his beam of light over to Sonny. The doctor immediately shook his head.

"The boy is gone." There was no doubt in his mind. The skull had been badly crushed, likely from the impact when the Ford struck the telephone pole. There was no way anyone could survive such blunt head trauma. "We need to get that driver to the hospital."

"Our problem is getting him out of this car," said Charlie.

The sheriff turned to the men gathered up on the track.

"Can we get some manpower down here?" he shouted.

Several of them were railroaders on the train. They had walked back to the wreck site once the powerful machine finally stopped. They were two section men and two brakemen, the plow master, the engineer, and the conductor. Down to a man they were visibly shaken, despite knowing there was nothing they could have done to prevent the wreck.

As they waited for the investigation that would likely occur once the passengers were removed from the Ford, the railroad crew offered what help they could. Some of them broke away from the crowd and waded down the bank when Foulkes asked for assistance. Bill Sawtell and the two Brownville constables were also helping. The men got a grip on the body of the car and heaved slowly until it rolled back onto its tires. Daniel Speed had now been in the car for almost two hours and was still unconscious.

Charlie Foulkes saw headlights driving along the embankment, coming toward them. It was a Brownville plow truck. He hurried over to meet the driver and ask for a crowbar. Back at the Ford he was able to pry open Danny's door. They would get to Sonny next. It was hard to tell in the twilight and the falling snow, but Speed seemed to be still breathing. Dr. Carde watched as the ambulance driver and his partner eased the body out from behind the steering wheel. Danny's jaw was visibly dislocated, and Carde assumed broken. Cuts caused by flying glass dotted his face, and his shirt and jacket were covered in icy blood that had coagulated. With several men holding the sides of the stretcher, they carried him up the embankment and down the tracks to the ambulance.

Dr. Carde followed. Neighbors who had gotten flashlights from their homes walked next to the doctor, their yellow beams lighting him the way. Carde climbed into the back of the ambulance. He would accompany Danny Speed to the hospital.

The Speeds were still new to Brownville, having made only a few acquaintances. From what the sheriff could understand when he queried bystanders, no one had felt comfortable enough to let Speed's wife know he had been in a terrible accident. With the storm still raging and every able man needed at the wreck site, he doubted anyone had gone to inform the Pomelows, either. But then, it wasn't their jobs.

"Go to Bill Thomas's house," Charles Foulkes instructed one of the constables. It was less than a mile away. "Tell Mrs. Speed what happened. Drive her to the hospital if she needs a ride."

They removed Sonny Pomelow next. The impact of the car striking the telephone pole had smashed the passenger door, caving it in. The boy would have to be pulled from the driver's side of the car. The men struggled to extract Sonny's legs from the collapsed metal. Dr. Carde had given instructions to have the body taken to Keniston Funeral Home where he would write his medical report later. Even the most medically unskilled bystanders could see that Sonny was dead. His bulky coat hid the fact that his chest had been crushed. But the fractured skull was evident by the drooping on one side of his head. Blood and fluid had issued from his nose and right ear and was now crystallized in the cold.

As he was taken from the car, both men and women standing up on the tracks were holding back sobs or were outwardly crying. A man passed two blankets forward and they wrapped the teenager in them. Neighbors watched as the body of Ray Pomelow Jr. was carried down the tracks and slid into the canvased back of the plow truck. They stood silent in the snow and cold until the red taillights disappeared from the crossing.

Deputy Sheriff Charles Foulkes did what his job demanded. A dull ache in his stomach, he climbed up the embankment and walked back to his car. Putting it in gear, he fought snowdrifts on the road as he drove to Stickney Hill. He parked in front of the Pomelow house, next door to the elementary school. He needed to inform Ray Sr. and Grace that their only son was dead.

◆

Again following the plow truck, the ambulance made its way over the narrow highway back to Milo, its light flashing and its siren breaking the stillness of early evening. Dr. Carde sat next to Speed in the back of the vehicle. The young man was still breathing. Ahead, the white road disappeared beneath the car to the rhythm of the wipers as they worked to clear snow. Carde had had a personal involvement with a train hitting a car. Nearly twenty years earlier, his wife and young son were struck at a grade crossing down near Burnham, where Mrs. Carde was visiting family. She had heard the whistle and saw the signal, but was unaware that it was a *Y* and two crossings. Albert was on duty at the hospital in Milo when he received word. Too far away to act quickly, he phoned a colleague to attend to his family until he could drive the fifty miles. It had been a long ride that September day in 1933, the longest of his life. He arrived to find that they were badly bruised and requiring stitches, but miraculously both wife and son were otherwise fine. The car was demolished.

When the Keniston ambulance pulled into the hospital driveway, another doctor and two nurses were waiting inside the front door. The small hospital at Milo had sixteen beds and was often a stopover for more serious injuries until they were transferred to the larger facilities and specialized surgeons down in Bangor. That would also be the case with Daniel Speed. The doctors on duty could work to stabilize his condition for now. With drifting snow blocking highways as it was, only a train would be able to take the patient on to Eastern

Maine General. And a plow train would likely need to clean the track first. The B&A passenger train No. 8 was due south from Millinocket that evening and had already been delayed by snow. It was likely that Daniel Speed would be put aboard No. 8 when it was able to reach Milo and continue on to Bangor. It was almost nine o'clock and he was still unconscious.

Exhausted, Dr. Albert Carde left his patient in good hands and walked down the street to his house, which was next door to Keniston Funeral Home. Someone would let him know when Sonny Pomelow was brought there.

MONDAY EVENING

And ere the early bedtime came
The white drift piled the window-frame,
And through the glass the clothes-line posts
Looked in like tall and sheeted ghosts.

—"Snow-Bound: A Winter Idyl"
by John Greenleaf Whittier

HOWARD JOHNSON'S AT THE KENNEBUNK EXIT

The northeaster had now lost most of its power. But the thirty-one inches of snow that had fallen overnight and all day were still engaging with strong winds in many parts of the state. Dozens of plows in towns along the turnpike, including some large state plows, were up and running. That afternoon two police officers had come into the still-packed Howard Johnson's to inform the southbound motorists that it was now possible to continue on their way, but only on Route 1. The others would need to wait a bit longer. North to Portland and beyond was still being managed, especially Route 1 as it left Brunswick and followed the ocean. That highway was not expected to be cleared until late that evening or the next day.

The visitors had fared well during their short but strenuous ordeal. Local churches and the Red Cross were still tending to those who had been transported into Kennebunk and given hot food and blankets. But upon hearing this newest announcement a couple dozen southbound drivers bolted for the restaurant's door in an effort to get back on the roadway ahead of the throng. Others, especially those with small children, opted for the safety of morning and daylight. Exhausted from the 24-hour workday, several of the Howard Johnson's staff drove home or got rides. They needed sleep before returning to serve breakfast to the remaining drivers. They would send them on their way well-fed.

Three waitresses and chef Arthur LeBlanc stayed a few hours longer to offer a light supper and then prepare for morning. LeBlanc would be back at dawn to start the muffins baking. He finally drove home to be with his wife and little girl, hoping to get some rest. He and manager Jimmy Ivers were again being praised all around for their skillful handling of a difficult situation.

THE CITY OF BATH AND VARNEY MILL ROAD

Hazel Tardiff had called her sister Evelyn to ask if the streets were being plowed. When Phil came into the kitchen from replenishing the fire in the furnace, she had already hung up the phone. He stood over the register enjoying the hardwood heat. Hazel put a slice of bread in the toaster. She already had a pan of milk warming on the stove. Phil went to the window and stared out at the road. Clouds of windblown snow from off the Kennebec River swept across the driveway in white curls.

"The plow ought to be along any minute," he said, his tone optimistic.

"How about some toast?" Hazel asked. When hers popped, she put two more slices of bread into the toaster.

"I expect they'll send their big Walter plow," Phil added. He pulled back a chair and sat at the table to wait. Hazel buttered the toast and put it on a plate in front of her husband. She shared her warm milk by pouring half into a second cup. She sat at the table across from him.

"I hope the plow don't wake you when it comes," Phil added. "That Walter makes a lot of noise."

Hazel smiled as she looked at her husband. All those nights in Portland when she crept into his hospital room and checked that he was still breathing, she had seen those qualities in his face that she hoped for in a future husband. That's why she made decision to put away her nurse's uniform and marry Phillip Tardiff. Now fate had brought them to Varney Mill Road.

"I was just on the phone to Evelyn," she said. "They're still trying to get the plows up and running. When they do, they'll have the whole city of Bath to clean up before they get to Varney Mill Road. We'll just have to deal with whatever comes."

Phil said nothing. After a moment, he reached across the table and touched Hazel's hand.

KENISTON FUNERAL HOME, MILO

Dr. Albert Carde had a short distance to walk from his house, which sat next door to the funeral home. The town was quiet, except for the sounds of taped laughter coming from the rare television set. Fresh snowed glistened beneath porch lights. It had stopped falling earlier and the wind had settled down, except for an occasional gust that pirouetted down the sidewalk. He stomped snow from his boots at the front door that had been left unlocked for him. Hearing his entry, a nighttime employee came from a back room, rubbing sleep from his eyes. He offered a weak smile.

"He's on the table," he said, and Carde nodded.

He stood for a minute at the door to the embalming room where someone would come by in the morning to prepare Sonny for the funeral service. It was such a loss at that age, and Carde had never gotten used to it. Mostly, as he removed the sheet from Sonny's body he thought of his own son and the train that had demolished the car he and his mother were riding in. Who plays that kind of poker with fate? Who holds the cards that determine which boy lives and which one dies?

He began the necessary work, as much as he disliked his job at moments like that. Each time he examined a part of Sonny's body he made a note. Within an hour, he was finished.

On the report he had written what he had mostly surmised earlier: *fractured skull, crushed left chest, fractured jaw and fractured left femur in mid-shaft. Death caused by fractured skull and crushed chest.*

Albert Carde covered the boy's body again with the sheet. He flicked off the light and closed the door behind him. He needed some sleep.

CAP MORRILL'S TAVERN

Jimmy was back behind the bar at Cap Morrill's that night. Those local patrons who could walk through snow to the tavern had arrived, bundled in coats and boots and ready for an evening of conversation about the storm. More than once they asked Jimmy and Pete to tell the story of Branch Lake, how the snow had trapped them in the Hillman for the night. Already a reporter from the *Bangor Daily News* had sat on a bar stool and asked his questions before leaving to write a story. It would appear in the newspaper the next day and Jimmy figured it would be July before customers stopped asking about their lake adventure.

Pete helped with washing glasses and pouring drinks. It was an evening of joviality, in the warmth and safety of the tavern. Customers wanted to know about the Hillman Minx, which was a conversation piece in itself. When did Pete plan to retrieve it?

"When the storm is over," Pete said. "I'll need to hire a plow truck."

"You should have taken your Rambler," one of the regulars teased Jimmy. The Nash Rambler was known for its diminutive size, being one of the first compact cars ever produced. "You could have driven home on top of the snow."

Laughter rocked the bar, a sound Jimmy liked hearing at Cap Morrill's. When they closed that night, he wiped down the bar top and mopped the floor. Then he ate a sandwich in the kitchen before falling dead tired into bed. Laddie was soon sleeping near his feet.

Earlier in the evening, Pete had tried again to find news of the missing Vickers. Finally, there was an update. The search party from the Italian village of Burgio had followed a mule track up the rugged slopes of Monte delle Rose. Near the top, driven into the earth, they had found the twin-motored plane. The pilot, ignorant of wind conditions on his route, had failed to gain the necessary altitude. The small plane had crashed and shattered into the mountaintop. Previous reports of survivors waving were incorrect. The crew of five, and the twenty-six women and children on their way to husbands and fathers in Kenya, had been killed instantly.

Pete couldn't put words to the sadness he felt for these strangers. Somehow, they linked him to his own history. He sat alone in the small living room and smoked his pipe before he went to bed that night. His mind was on his fellow British soldiers in Africa, and not on his and Jimmy's adventure at Branch Lake.

PORTSMOUTH

It was now the second worst day of Ellie Haigh's life. With phone lines down in many places and emergencies to contend with, Wiggin Funeral Home in New Hampshire had not yet been in touch with Maine concerning the body of James Haigh. For hours, no one in New Hampshire had been told where Jimmy was, much less discussed details. Thus, when the call came earlier from the funeral home up in Thomaston, asking the Haigh family for particulars in returning Jimmy's body to New Hampshire, Ellie was not prepared. Rather than giving them the phone number for Wiggin and letting professionals deal with it, she asked the director why her husband could not come home immediately. Because of the storm, she was told. And yet, her heart broken, she begged. Please, wasn't there some way so he could be with his family? There wasn't.

Hours later, with all the well-meaning guests gone, she had fallen into bed. When she finally fell asleep next to Barbara Ann, she slept soundly.

PLEASANT POINT

Alice Davis was now Harland's widow and not his wife of fourteen months. Her husband had been the last person to speak to Ellie's husband, and yet the two women had only a vague notion that the other existed. They did not know then that they were both the widowed mothers of young girls. They only knew and cared about their own loss and their own family's grief. It was an unwanted kinship that the winter storm and the ocean had dealt them. Their husbands had shared with each other the last words they spoke on earth.

Alice woke many times in the night, once from a dream that Harland was waiting for her at the movie theater in Rockland, a smile on his face. Each time she managed to fall back into a slumber, the dreams were waiting for her.

HULLS COVE AT BAR HARBOR

Paul Delaney was losing his sense of time, despite glancing at his watch every hour. The boredom he had felt at first was blending into loneliness. He had napped, sometimes thinking it was for ten minutes, then checking his watch to see that it had been for an hour. Once, he woke shaken, certain he heard his little sister calling from her bedroom next door. "Paulie, are you all right?" It was her pet name for him. He sat up quickly, his limbs cramped. In the darkness he had no idea where he was. "I'm okay, Sarah," he whispered.

Wide awake, he felt foolish when memory came back to him. He had gone to Bar Harbor to take Mona to a movie. The snowstorm. He should have left earlier. He slid down Ireson Hill and into a ditch on his way back to the naval station. His dream had brought back the memory of the car accident he had in high school, a minor collision. When the police arrived at the house to tell his parents, Sarah was listening from the top of the stairs. She thought her brother had died. When Paul got home a few hours later, released from the hospital with just bruises, she ran to him, clutching his hands. "Paulie, are you all right?"

He would call his family as soon as he got out of this mess. He would send Sarah a small gift from Maine. A doll or a teddy bear. He hadn't kept in touch as he promised. His wristwatch told him three o'clock. He was certain that meant early morning. He had been in the darkness of the car, smothered in snow, for thirty hours.

During that day, as local residents snowshoed down Ireson Hill on their way to town for supplies, they had no way of knowing that beneath the snow now drifted to ten feet high in the ditch at the foot of the hill was a young sailor buried alive in his car.

MILO HOSPITAL

Frances Speed had taken her two sons, the baby and the toddler, to her sister's house. Too shocked to cry, she had ridden with the constable over snowdrifts to the small hospital at Milo. As the hours passed, she refused to leave her husband's room. With his vital signs unstable, and in his unconscious state, he had been placed on the critical list. When the doctors left Danny alone for a few minutes, Frances stood by the bedside. She held his hand and talked to him, hoping he could hear her. She spoke of their sons who needed him, the home they hoped to buy one day, his worried parents.

"Danny, you can't leave us," she whispered.

Frances also knew that if Danny recovered, he would have to learn that Sonny Pomelow had died in the wreck. She knew her husband's heart. He would be inconsolable. She remembered the boy shoveling the front porch and yard for his grandparents, a quiet boy with a shy smile. Danny liked young Sonny. Now the boy was at Keniston's Funeral Home.

Midnight came and went. At three o'clock the B&A passenger train No. 8 had a track headed south that had been cleared by a plow train. The Keniston ambulance delivered the patient from the hospital to the waiting train. As Frances and one of the Milo doctors sat on each side of Danny's body, No. 8 rocked back and forth along the track and belched fumes all the way to Bangor.

It was 4:30 that morning when drivers for the Clark-Mitchell ambulance heard the whistle blowing and saw clouds of smoke appear in the predawn. A doctor and nurse from Eastern Maine General were waiting for the patient to arrive. Two police officers were also there. And a reporter for the local paper had turned up to gather facts for a story. His photographer snapped photos as Danny's stretcher was lifted down from the train and put into the back of the ambulance. With Frances still at his side, the car sped to the hospital over roads plowed especially for the emergency. Frances had no idea if her young husband was going to live. Daniel Speed had been unconscious for more than fourteen hours.

THE MISSING CAT

By bedtime that night there were still no paw tracks on the front porch. With his boot, Bill cleared a small space in the snow near the door. He carried a cardboard box from the kitchen and placed it there. Inside, he had laid an old flannel shirt for Snooky to curl up on should he return. In a corner of the box was a bowl with leftover salmon from supper.

It was just before dawn that Bill rose to use the bathroom. When he opened the door and peered out onto the porch, he saw that wind had filled his box with snow, burying the shirt and the bowl of salmon. There was no sign that Snooky had been there.

PART FOUR

FEBRUARY 19, TUESDAY

The sled and traveller stopped, the courier's feet
Delayed, all friends shut out, the housemates sit
Around the radiant fireplace, enclosed
In a tumultuous privacy of storm.

—"The Snow-Storm" by Ralph Waldo Emerson

DIGGING OUT: THE AFTERMATH

Towns with stricken plows were still incapacitated. The more rural ones had been isolated for hours to any incoming traffic, and at least one had used dynamite to unblock the twenty-foot drift that barricaded a hilltop. But many larger towns had cleared away enough snow by Tuesday morning that modest traffic was moving again in business districts, especially in those places where the wind had stopped blowing. The Maine Turnpike and Route 1 were slowly being reopened. Citizens took to driveways and streets with shovels, hoping to make way for firetrucks, ambulances, and milk deliveries.

Heart attacks from overexertion in the snow, whether walking, pushing an automobile, or shoveling, are expected with every deep snowfall regardless of the warnings put out by the weather bureau. Many shovelers are not in the best physical condition and are unused to heavy exercise. And 1952 was during the period when cigarette smoking by adults would reach an all-time peak of 45 percent of the population. Lifting hundreds of pounds of snow puts a strain on the body. Added to this, cold weather can increase blood pressure and interrupt blood flow to the heart.

If our Neolithic ancestors moved snow, it's likely they used shovels made from the same material as the ones that moved their rocks and soil. Archaeologists have found Stone Age shovels made from the bones of oxen shoulder blades. Not deserving of the great praise accorded to the wheel, a shovel nonetheless has played an important role in the rise of civilization. Before mechanization, it was the go-to implement. It dug canals for irrigation and built bridges. Many homes stood over foundations excavated by hand. Metals and minerals had to be mined. But using a shovel also helps to speed up a couple dozen heart attacks for each snowstorm. Now the Boston newspapers were already listing twenty dead in New England from heart attacks alone, one man as young as forty-three. And the shoveling out was just beginning.

A snowstorm often does diverse destruction beyond the taking of lives, both human and animal. Newspapers and radio stations in Maine were busy reporting damages from those heavily hit communities around the state. Automobile wrecks and other accidents had been numerous, and property destruction was widespread. A notable story concerned businessman and future Maine governor Burton M. Cross. The owner of a large florist business, Cross had spent the day estimating the extensive damages done to his six greenhouses. The largest had collapsed under the weight of the snow, destroying 6,000 carnations.

Local heroes rose to face a myriad of difficulties. Residents read about the father whose toddler overturned a percolator full of hot coffee on himself and was badly scalded. He found salve for the baby at his neighbor's and then plodded seven miles in deep snow to telephone for help. An alerted state trooper made his way through a mile and a half of knee-deep snow to carry the boy back to his car and drive him to the hospital.

With rural roads unplowed, a farm burned to the ground as neighbors frantically tried to save it with an old-fashioned brigade of water buckets. By the time a fire truck reached the scene, following behind a snowplow, it was too late. The uninsured house, shed, garage, and barn were charred shells. Stored in the barn had been fifty tons of hay.

Gale winds blew a barn door off its hinges, hitting a young farmer in the head and knocking him unconscious. He was recovering after being hospitalized with a serious head injury.

In one town, fire broke out in the brooder pen of a chicken farm. With firetrucks unable to reach the building, over a thousand baby chicks died. That same fire also destroyed a ton of grain and hay.

A plow struck a buried automobile with four exhausted and cold passengers huddling inside, five feet below the surface of the snow.

A farmer's wife, expecting her ninth child, was pulled by sled a mile over snowy fields to the home of a doctor while her husband stayed behind to babysit the other eight children. The doctor delivered a healthy baby girl.

A bus from the Maine Central Transportation Company that had gone missing on Monday was found buried beneath a twenty-foot snowdrift, the driver and three women passengers inside still doing fine.

Two teenaged girls missing overnight were found trembling with cold in the waiting room at an empty bus station.

Two snowplow operators had been overcome by carbon monoxide fumes inside their cab and were recovering, following medical advice via a doctor's telephone call.

A diligent postmistress put on snowshoes and trekked for two hours from her rural home to reach the post office. When she discovered that no mail had arrived to be delivered, she put her snowshoes back on and trekked home again.

On the Maine Turnpike a man got out of his vehicle in whiteout conditions to assess the damage of a minor collision and was struck and nearly killed by a passing car.

With thousands of babies and toddlers to feed in several states, milk became a precious liquid. A father, hoping to carry a bottle from his stalled car home to his infant, fell on Route 1 and was seriously cut by broken glass.

A doctor and his wife loaded a sled with milk and hauled it three miles to the hospital where he worked. Lucky patrons in Portland, however, had U.S. Marine reserves delivering bottles along some streets. But one dairy farmer, unable to get his product to market, fed the milk to his hens.

When two trucks filled with thousands of eggs were stuck in snowdrifts, heaters were found by local residents and brought to the rescue to keep the eggs from freezing.

Sick children were taken to the hospital with the help of policemen and neighbors. A boy with pneumonia was carried a mile atop the snow on a stretcher and put into a police car.

Four hundred men, women, and children who were still trapped in autos along Route 1, most of them fans returning from the Ice Follies, were now desperate. Late Monday night, the Boston and Maine Railroad sent two passenger cars following behind a plow train to the rescue. Over thirty sailors stationed nearby formed a human chain from Route 1 down an embankment to the tracks and guided the tired travelers into the passenger cars, some having waded three miles through snow.

All over New England thirty-foot seawalls demolished small ships and boats, along with lookout towers and wharfs, as waves crashed into coastal residences doing great damage to the structures.

THE UPBEAT STORIES

Newspapers were also busy using ink in order to relay a lighter side of events for those readers who enjoyed special interest stories.

Up north, where Aroostook County had just recovered from its own heavy snowfall, there were no storm-related deaths or major accidents with just ten inches of accumulation. Trouble was expected in Houlton, however, when two firetrucks went racing to answer an alarm call coming from the box inside the local hospital's main entrance. A fire could mean disaster for dozens of trapped patients. Upon arrival, they found a confused patient trying to push a stamped envelope into the fire alarm box.

At least one celebrity was caught in the blizzard, John D. M. Hamilton, a veteran politician and lawyer who had gained national fame for representing Harry Gold, the chemist convicted of being a courier of American secrets to the Soviets. Gold had just testified in the Julius and Ethel Rosenberg trial and the Rosenbergs were at Sing Sing awaiting execution in the electric chair.

At sixty years old, John Hamilton was now the eastern campaign manager for presidential hopeful Robert Taft, son of the former president, and had come to Maine for political meetings. His train was met in Boston by two young law students and soon all three men were stuck in snow at the Biddeford exit, less than a mile from Charles Voyer. When dawn broke, Hamilton and his chauffeurs trekked to a farmhouse where the politician got some sleep while he waited for roads to open back up. A reporter tracked him down before he caught a train out of Portland's Union Station on Monday night. "If Taft can snow under his opponent like Maine has snowed me under, I'll be perfectly satisfied," Hamilton was quoted as saying, well aware that his remarks would make Maine's largest newspaper the next day.[10]

Anxious to be released after three days of incarceration, two jailed occupants agreed to plead guilty to intoxication and vagrancy during a telephone call to the judge who couldn't get out of his blocked driveway to hear their cases.

One city official referred to the coins in parking meters that were buried in snow as the town's "frozen assets."

A grocery store worker who was snowshoeing to work suddenly realized he was walking next to telephone wires. When the tip of his snowshoe struck against metal, he discovered a city bus that was buried in the snow beneath him.

A bride was determined to keep the wedding date with her young soldier-groom, home on leave to marry her. She snowshoed a mile to the church, meeting up with her groom who was also on snowshoes.

While residents were required by town ordinances to keep fire hydrants and sidewalks cleared, there was the ambitious lad who kept shoveling out the parking meter in front of his house, even though a car would need to be airlifted to reach it.

A worker who turned up at his place of employment sat in the boiler room at midnight to eat his long-delayed supper. When he noticed a skunk at his feet, he gingerly gave the animal part of his sandwich and then some water to drink. Thankful, it waddled off into the storm without spraying him.

A potential store robber, certain the tempest was on his side, fell from the roof of the business he hoped to rob and lay with a broken leg until police drove him to a hospital, and then on to jail with his new leg cast.

In the town of Norway, known as the "Snowshoe Capital of the World," the company that had once provided polar explorers Admiral Byrd and Admiral Peary with skis, sleds, and snowshoes, shut down production "due to heavy snow."

In Portland, where abandoned cars were like leaves scattered on sidewalks, one driver left his Ford on an icy bridge with a crude sign showing in the back window: FOR SALE!

Residents of one town organized a "Shoveling Bee" and opened a quarter mile of road so that a truck could deliver its cargo of heating oil.

For added comic relief, a scene unfolded on a street in downtown Portland that could have originated in a horror film. In front of a once-busy department store lay what appeared to be a naked, dead body. Passersby were so startled that many screamed and some fled. Police called to the scene found an undressed manikin that had blown onto the sidewalk when sixty-five-mile-per-hour winds smashed through a display window.

And, as expected, there were still the occasional complaints coming in, usually from those people in various towns who were famous for grievances. Why hadn't the mail been delivered yet? Why wasn't the daily newspaper on the front steps? Why weren't the streets cleared? One man, on the first evening of the storm, called his town office to complain that he was not just starving, he was also out of cigarettes.

It was the news of loss that held the state's greatest attention. Two of the twenty men being listed in New England as deceased from exertion in the snow were

Charles Voyer and George Aspey, who had died in Maine. But those unanticipated deaths hit communities the hardest. The state was slowly learning the stories. Two young fishermen had died in the ocean off Port Clyde on Sunday. And a plow train had struck an automobile in Brownville the day before. With whiteout conditions blanketing the railroad tracks, the driver had poor visibility. He had finally regained consciousness that morning at a Bangor hospital, but was still on the critical list. His passenger, a high school boy just fifteen years old, had died in the collision.

For the most part, New Englanders held on without complaints. Mainers shoveled out and hunkered down. Northern ravens were again circling over town dumps in search of garbage blown by the wind. Flocks of grosbeaks and chickadees fed on the bread and seeds put out by residents now that bird feeders could be filled again.

At the Howard Johnson's restaurant near the Kennebunk exit, dozens of cars were slowly pulling out of the parking lot, those remaining drivers getting back onto the road after enjoying one of Arthur LeBlanc's fresh breakfast muffins and a cup of hot coffee. The day before, those travelers had taken up a collection for LeBlanc, who was once a dishwasher at the restaurant, a generous tip to be shared with his loyal employees who worked through the storm to make them all comfortable.

Despite the uplifted moods, weathermen were predicting that another northeaster could be on the way.

PLEASANT POINT

It was the first time Alice had seen sunshine since early Sunday morning, before the storm reached the coast of Maine. Now the sun rose with daybreak over the water, peaceful and calm, breaching yellow through patches of gray cloud. She hated the feeling she had just then, that the world was going on without Harland in it. The world was acting as if no great loss had even occurred. It would take the rest of the day for Pleasant Point and nearby villages to dig themselves out, but the worst of the storm had passed. Wind wafting in from the water still riled the banks of snow. But life was stirring again. Folks were walking down to the store for milk and bread, stopping to talk when neighbor met neighbor. The conversation was still about the loss, one of their own, another fisherman claimed by the sea. The orange derrick on Harland's wharf, inches of snow still coating the platform, was a poignant reminder.

The evening before, Alice Davis had given Martha the bag of clothing she had chosen for Harland to wear in the casket. As a favor to the Davis family, Martha had offered to drive it up to the funeral home in Thomaston as soon as the highway was passable and safe. There would be no wake, given the bad roads and now the uncertainty of another possible storm on the way. A funeral service for Harland Creamer Davis would be held in Thomaston on Thursday. The burial, as planned, would be in the spring. When Alice saw her husband again, he would be dressed in the dark blue suit he had worn on his wedding day. Before folding and packing the suit into the paper bag, she had slipped a photo of her and Carolyn into the breast pocket. On the back she wrote, "We love you."

Spring seemed a long way off, with its sounds of ice dripping from the eaves of houses and the calls of returning warblers. By the time the silver poplar down near Harland's wharf was green with leaves again, maybe she would have a better idea what to do with the rest of her life. The baby was due that summer. Alice French Davis needed a plan. So far, there was none on the horizon.

ALONG ROUTE 1

A black Chevy Styleline deluxe station wagon with New Hampshire plates stood out on Route 1 as it rolled slowly north over the white road to Thomaston. A year old, its smooth curves of chrome and stainless trim sparkled silver in the morning sun. In small towns and villages, locals pay close attention to the sound of a siren. It means one of their own is in trouble. And they pay deep respect to the appearance of a hearse. It means one of their own has died. There was no doubt that the glossy black station wagon that rode quietly through villages was a hearse. The Cadillac that Wiggin transported caskets to and from wakes and burial services would be used later. But for this task, the station wagon was the better choice.

When the driver from the funeral home pulled over in Brunswick to use the restroom and buy a coffee at the diner, two teenaged boys came out to inspect the car. Boys who loved cars knew that this one was all steel, and not the wooden body that Chevy had been manufacturing in previous years. Questioned while getting his coffee, the driver admitted to the clerk and a few locals who were peering fixedly out the window that he had, indeed, come for the New Hampshire man who died off Port Clyde on Sunday. No one in the diner spoke as they watched the station wagon pull back onto the road and continue north.

THE CITY OF BATH

Bath was working overtime to regain its composure. The navy had sent a huge plow from the Brunswick naval base to help remove the snow. Some of it was scooped up by loaders and dumped into trucks to carry off to the Kennebec River. But there was much work still to be done, especially in the outlying rural areas where roads remained clogged with four and five feet of snow. Most of the crews who had been on the job since early Sunday afternoon had finally gotten some rest and were back to work. Better news was that Stanley Peterson, having left the Howard Johnson's restaurant as soon as daylight broke, had arrived in town with his bus full of exhausted Ice Follies fans. They were still dressed for the Sunday matinee, many of the females wearing skirts and nylons, the men in slacks and dress shoes inside galoshes. The bus dropped them off at the courthouse on the corner of High and Central. From there, some of them had a half mile to walk in deep snow to finally reach home.

As parking meters began to reappear from beneath deep snow, Bath was already buzzing with other news. Folks had learned the day before that two large tankers—each was the size of two football fields—had broken in half off the coast of Massachusetts, the SS *Fort Mercer* and the SS *Pendleton*. The coast guard there had launched individual rescue attempts in sixty-foot-high waves to save as many of the stranded crewmen as they could. Bernard Webber, a petty officer 3rd class, had been at the helm of the CG-36500, a thirty-six-foot lifeboat that sped into the raging sea, headed toward the *Pendleton*. The boat returned with thirty-two rescued survivors who had been stranded on the stern section of the tanker. Sadly, nine of the crew on the *Pendleton* were lost, and four from the *Fort Mercer*. But Bernie Webber was on his way to becoming a national hero.

It so happened that Webber was the grandson of the late Ruel Knight and his wife Margaret, whose house was on Washington Street. Annie Knight, his mother, had been born and raised in town. Many in Bath knew Bernie well since the young man often summered there with his parents. Margaret Knight's phone line had been humming all night and morning. She and other relatives were thankful that Bernie was alive since the odds had not been with him in those seas. It was likely that Bath, and the entire country, would be discussing the details of the courageous rescue for weeks to come.

VARNEY MILL ROAD

Out on Varney Mill Road, Hazel Tardiff stayed in bed that morning, warm beneath the homemade quilt she had bought at the craft fair. It reminded her of those childhood days on Isle au Haut, when she and her sister would bring heated bricks or rocks to bed during the cold winters. She heard Phil down in the kitchen, the sounds of the fridge door opening and closing. Then the faucet water running. Pans rattling. The toaster popping with its metallic ping. David's voice broke in, asking about firewood he was to carry from the barn. Her two girls left their bedroom across the hallway. Hazel heard their footfalls as they descended the stairs, whispering to each other and giggling. This was the sound of the Tardiff family, *her* family, and it gave her comfort knowing they could function well enough without her. But she should be up and getting breakfast instead of Phil, who had insisted she stay in bed. She didn't resist. She had been awakened at dawn with contractions that had not lasted long or been too painful. But she felt quite sure her labor would begin in earnest either that day or, heaven forbid, during the coming night. It was time to check into the hospital.

Phil had just put fried eggs on Mary Lou's plate when Hazel appeared in the kitchen. He smiled when he saw her. Phyllis went to hug her mother.

"Want me to get you a cup of coffee, Mama?" she asked.

"Maybe a small glass of orange juice," Hazel said. She sat in the chair Phil pulled out for her at the end of the table.

"I see the road hasn't been plowed," she said. She knew from her sister the situation in Bath and wasn't surprised. Phil shook his head.

"I doubt they'll make it out here today," he said. "Downtown is still a mess."

There was no more pretending. Hazel looked at the anxious faces of the children as they waited for her to respond.

"I'll call Dr. Hamilton," she said. "We'll figure out what to do. Meantime, I don't want any of you worrying. Just enjoy another day with no school."

When this brought a round of smiles, Hazel reached for her orange juice.

THE MARDEN FARM, SACO EXIT

The Saco Police Department had telephoned Harriet Voyer the day before to inform her of the bad news, that her husband Charles had suffered a heart attack after exiting his car on the turnpike and plodding over snowy drifts to reach the Marden farm. With Buxton Road now open, a hearse from Dennett-Craig Funeral Home in Saco drove slowly out to the farm and backed up to the front door.

The other stranded guests, including the young couple who had tended to Voyer during the night he spent in his car, had already walked back to their plowed-out vehicles and gone on their way. Dennett-Craig would deliver Voyer's body to a Portland funeral home once Mrs. Voyer gave them instructions. Now, after spending a night in the back house on the Marden's farm, Charles Voyer was finally on the last leg of his trip home to South Portland, less than fifteen miles away.

VIRGINIA CLAY HAMILTON

Dr. Hamilton was sitting at her desk when the phone rang. As she answered, she swept away ashes that toppled from her cigarette onto the desktop. A local journalist had put in the *Bath Daily Times* a picture of the cherry desk in Virginia's office on Washington Street. According to the article, the desk had once belonged to Henry Clay, Kentucky's famed statesman known as "The Great Compromiser." It had thirty-five pigeonholes, some tall enough for big ledgers, others so small a bottle of ink would fit snugly. The article about the desk caused quite a stir in town. Dr. Hamilton stood out like a movie star in Bath, well-known for the perpetual cigarette burning between her fingers or dangling from her lip if her hands were busy. She was rarely seen without one even though she once lectured a church group that the body was a temple, in need of both spiritual and physical upkeep. The local women liked her despite the smoke curling into the air as the doctor leaned forward on her elbows to listen closely to their words. More importantly, they trusted her to deliver their babies.

Hamilton had short, naturally blond hair that she wore in a bob. There were whispered jokes about her style of dress. She had been seen more than once wearing bobby socks with high-heeled shoes. But most often, and unlike local women, Virginia wore pants with Buster Brown shoes and socks. Skirts or dresses she reserved for special or formal occasions. When she turned up at the hospital in a long leopard coat everyone assumed it was real fur. It was. Her father was a Kentucky state senator and her early upbringing in Lexington was one of privilege. She was thirty-seven years old when she married for the first time, to fellow Kentuckian Boyd Bailey. The following year they settled in Bath, where Boyd would later become Maine's assistant attorney general and Virginia would deliver a lot of babies.

It would surprise her patients if they knew that on Virginia Hamilton's wedding day, which had taken place in New York City a dozen years earlier, she was dressed in the taste of her upbringing. She wanted no part of a large wedding with a fairytale gown. But her woolen suit was a Molyneux, by one of the top British designers of the day who had a successful salon in Paris.[11] Her blouse was made of fine Belgium silk, and her shoulder bouquet was a mass of roses and lilies of the valley. A celebratory breakfast had been at one of the city's most prestigious hotels. This was not the same doctor well-known for shopping at the Grants store, or picking up lobster rolls and French fries from Sam's restaurant on the corner of Middle and Pine Streets.

But Virginia was not the type of woman to rest on her inherited laurels. Her concern was medicine, not the fashions of the day. She had graduated from Cornell University Medical College and interned at Bellevue Hospital. As a young doctor she had done research on population in Sweden and Norway. She would tell friends that her one regret was turning down an opportunity to work with Margaret Sanger. Now, the local barber cut her hair, which she washed with a bar of Ivory soap. A dash of lipstick was her idea of makeup for a special event. And her favorite entertainment was a nightly cocktail with her husband, a martini before dinner one night, a Manhattan the next night.

It was Hazel Tardiff on the phone. She was having irregular contractions since she woke that morning. Their country road was still unplowed, making passage by automobile impossible. Her husband had called the city council and asked Rodney Ross, the chairman, if something might be done. Ross was calling the Tardiffs back. But city officials were at their wit's end just trying to get downtown Bath moving again. No other town in the state had suffered so many incapacitated snowplows as Bath.

"Let me phone you back," Virginia said.

When she got the dial tone, she called the highway department. This could be an emergency situation, she informed them. Her patient was about to have a baby. Was there any way to get a plow out to Varney Mill Road?

"If Jesus himself called for a plow out to Varney Mill, we'd have to tell him no," an exasperated male voice told her. The doctor hung up and called Hazel back.

"We'll figure this out," she told her patient. "In the meantime, you stay calm."

PORTSMOUTH, NEW HAMPSHIRE

It was just after breakfast when Earle Sanders came for the clothes. As Bubbles watched, Ellie Haigh sorted through the suits hanging in her bedroom closet. It wasn't as if Jimmy had a dozen to choose from. Like most men who worked hands-on in the fishing industry, wearing a suit was not a habit but a necessity.

"The dark gray?" Ellie asked Barbara Ann, who was sitting on the end of the bed. "I think he wore it last for cousin Tom's wedding a couple years ago. It should still fit." *Get through this, get through this, get through this* was the single thought running through Ellie Haigh's mind. She had to think of her daughter.

"Maybe with the light blue shirt?" Barbie asked and pointed to where her father's shirts hung. She was doing her best to be part of the decision-making so that her mother wouldn't feel alone. So was Bubbles.

"Good choice," Bubbles said. "Now what about a tie?"

"He liked the one I bought him for Christmas," said Barbie. "It's dark blue with gray diamonds on it. He never wore it yet."

Ellie and Bubbles exchanged a quick look.

Get through this, get through this.

Bubbles quickly took the suit from Ellie's hands and then the light blue shirt. She waited as Ellie sorted through the neckties that hung from a clothes hanger at the back of the closet.

"This looks like it," Ellie said, and pulled a dark blue one away from the others. She handed it to Bubbles, who forced a smile.

"I'll get this stuff down to Earle," Bubbles said.

Earle Sanders was waiting in the kitchen for the clothes. Wiggin Funeral Home had called to report that a car was on its way up to Maine since Route 1 was now open. It was time to bring Jimmy Haigh back to New Hampshire. When the driver returned, employees at Wiggin would see that Jimmy was dressed and then brought to 55 Gates Street for the wake that early evening. The street had been cleared of snow and it seemed that the city of Portsmouth was getting back to business as usual.

HULLS COVE

It was 9:00 A.M. and Paul Delaney had been in the buried car for thirty-eight hours. He was doing his best to keep track of time by checking his watch to the flare of his lighter and making a mental note of the hours passing. It was morning now if his calculations were correct. At least the cigarette lighter broke the darkness now and then and allowed his eyes to focus on the interior, the radio knobs, the steering wheel, the silver door handles, the white snow encasing the car. It reconnected him to the world. Otherwise, when he closed the lighter, it all disappeared into darkness, and the absence of sound. Surely they would find him soon. How long can a storm last? He was overwhelmed just then with the image of his mother's smiling face.

That's when Paul Delaney realized that tears were pouring from his eyes. He knew the story well. His parents, Eugene and Marie, had married in 1927, when the country was still enjoying the rapidly expanding economy of the Roaring Twenties. His mother had been an office clerk before her marriage. His father worked for a shipping and freight company. They saw only bright future ahead as they saved to buy that first home and begin a family. A year after they signed the bank loan, the stock market came crashing down. Wall Street spiraled into a panic as millions of investors were wiped out, some committing suicide rather than face the consequences. The Delaneys lost their home to the bank and were forced to move in with Eugene's widowed mother. Paul Vincent was their second son, born two years later in 1931. The sudden wave of remorse overwhelmed him. He made a vow to be a better son.

Seaman 1st Class Paul V. Delaney wiped the tears on his coat sleeve and crawled into the back seat, emotionally exhausted. He was beginning to sleep more than he was awake.

WINTER HARBOR NAVAL STATION

Jeffrey had expected Paul back with his car on Sunday night. Now it was Tuesday. But he knew that all of coastal Maine had been buried in snow since Sunday afternoon and nothing was moving on the roads. It was likely Paul had holed up with friends in Bar Harbor until the storm passed and highways were functioning again. Not everyone owned a telephone, but a phone call to the front office at the station to let Jeff know details would have been nice. He had to assume that his friend had no access to a phone. But if Paul didn't turn up that night or the next morning, he would have no choice but to let his commanding officer know. As it was, he was covering for Paul who had only the weekend off. He told himself not to worry. Delaney was a New Yorker. He knew how to take care of himself.

A MISSION ON SNOWSHOES

A taxicab with snow chains on its back tires was able to deliver Virginia Hamilton to the front door of Bath Memorial. She had one thought in mind and it was Hazel Tardiff, her patient out on Varney Mill Road. Hazel was the kind of woman who was soft-spoken and unassuming. She had also trained as a nurse, so Virginia knew that if Hazel felt it was time to make plans, it was time. She was about to telephone the police station to ask what might be done when Rodney Ross called the hospital. Ross was not only chairman of the city council, he owned a summer home on Varney Mill Road and knew the Tardiffs well. Phil had contacted him about Hazel's situation.

"If she can't make it to town," said Virginia, "then I need to go to her. I'll take a nurse with me. If the baby comes this afternoon, I'll be there to deliver it."

"There is no way that road is getting plowed today," Ross said. "But I borrowed Arthur Avery's car. I can get us as close to the Tardiff house as possible. We'll have to snowshoe the rest of the way."

Avery was the highway superintendent whose Chevy had been buried in snow at the foot of Winter Hill. Now free, it was being pressed back into service.

"Give me fifteen minutes," Dr. Hamilton told him. "Bernice Brawn and I will be waiting for you at the front door." Brawn was a superb RN, and Virginia knew she was the kind of Maine woman not intimidated by a pair of snowshoes.

When Rodney Ross drove up to the hospital entrance, he had three pairs of snowshoes in the trunk of Arthur Avery's car. Hamilton tossed her medical bag onto the front seat and got in next to it. Nurse Brawn sat in the backseat. Snow thumping the undercarriage, the Chevy left city limits and made its way along Old Brunswick Road before turning onto Ridge Road. This was the same route the 3.5-ton Walter plow had taken on Sunday night before it became wedged in snow. At Lover's Retreat Road, they came upon the plow, still stuck in the drift, its yellow paint showing through the snow in patches.

"This is as far as we get," Ross said, parking the Chevy just beyond the stalled plow. It was still over two miles to the Tardiff house. He opened the trunk of the car and tossed the sets of snowshoes down onto the snow.

"It's been a lot of years," Virginia Hamilton said. She was almost fifty-two years old. Those daily cigarettes wouldn't help. But she never doubted what her duty as a doctor entailed, and this was just an unusual facet of the medical profession.

Bernice Brawn occasionally snowshoed with her children and, like Rodney Ross, was more accustomed to them. But it had been years since any of the three had worn the contraptions. As they were strapping the shoes onto their boots, a truck from the highway department rolled up to the Walter plow. The plan was to get the big monster back on the job.

"If you get it out," Ross instructed the workers, "open the road to the Tardiff home. We might need it."

The three snowshoers set off down Lover's Retreat Road, atop four feet of drifted snow. There were a lot of little hills to cover in the two miles leading to the Tardiff home.

"What if the baby isn't ready to come this afternoon?" Ross asked the doctor.

"We should bring Hazel to the hospital," Virginia said. "The Tardiffs aren't prepared for a home birth."

"Do you know if they have a sled?" Ross asked.

Neither woman knew, and it was too late now to ask.

Alderman Joe Robinson's house sat on Varney Mill and was on their way. As Virginia and Bernice waited on the front lawn, Ross knocked loudly on the alderman's door.

"Joe, do you have a sled or toboggan I might borrow?" Ross asked. "Phil Tardiff's wife is due to have her baby and we might have to bring her to the hospital."

It wasn't like it happened every day, but Robinson didn't hesitate.

"Hang on," he said. "I got one on the back porch."

Neighbors along the road looked out of their windows to a strange sight: councilman Rodney Ross was pulling a toboggan and followed by Dr. Virginia Hamilton. The doctor was wearing a long brown coat with a thick fur collar and carrying her medical bag. Nurse Bernice Brawn walked behind. And all of them were on snowshoes. Seeing this, three neighbors on that road left their homes and joined the caravan in case their help would be needed. Now added to the group was Jim Gillies, his wife Laura, and Carl Dearborn. The two Dearborn dogs, tails wagging, decided to get in line and follow the excitement over the snow. Laura Gillies had grabbed her camera. Otherwise, who would believe it?

AT THE TARDIFF HOME

Taking the mode of transportation into consideration, and the hilly terrain, Phillip Tardiff told his family to expect the small rescue team to appear within two hours.

"Here they come!" Mary Lou shouted. She had been waiting at the window, watching the road. Excited, the other kids rushed to see.

"They're pulling a toboggan!" David added.

Phil stubbed his cigarette in the ashtray on the kitchen table and opened the front door. He stood waiting as the snowshoers made their way over the

snowed-in lawn, past the big Buick, to the porch steps. The kids watched as
the doctor unstrapped her snowshoes. Bernice Brawn stood hers up in the
snow, next to Ross's.

"How's this for a house call?" Virginia joked to the children. She stepped
past them and into the warm kitchen. Phyllis had hot coffee in the percolator
waiting for the guests and now they poured cups. Mary Lou put sugar and a
creamer of milk on the kitchen table.

Hazel was waiting in the living room.

"Do I need to deliver a baby here, or at the hospital?" Dr. Hamilton asked.

"It's getting closer," said Hazel, smiling. "We can go up to my bedroom so
you can examine me."

The doctor nodded. Delivering a baby at the Tardiff home was certainly
doable, especially since she had a good nurse with her in Bernice Brawn. But
nothing could beat the efficiency and sterility of a hospital in an emergency.
Phil followed the two women up the stairs.

Mary Lou and Phyllis cut the blueberry nutmeg cake Hazel had made for
the family to enjoy in her absence. They were passing plates of it to their guests
when Dr. Hamilton reappeared in the kitchen doorway. She paused to light
a cigarette and then accept a cup of coffee from Mary Lou.

"Hazel and I think the best decision is to go to the hospital now," she said.
"Judging by her dilation, I think the baby will be born this evening. She's
getting a few things packed."

Rodney Ross put his cup in the kitchen sink.

"Get me a couple of thick blankets," he said to the girls, "so it will be softer
for her on the toboggan."

Phil appeared carrying a small suitcase.

"Help your mother on with her coat," he told David, who grabbed a winter
coat from the rack near the door. Mary Lou found mittens and a scarf.

When Hazel came into the kitchen, she was all smiles, assuring the children
that things would be fine and not to worry.

"Don't forget the meatloaf in the refrigerator," she told Phyllis as she pulled
on her boots. "Bake it an hour at 375 degrees."

When she was bundled up warmly, Phil followed her out the door where Ross had the toboggan waiting. He helped her settle down onto the folded quilts, then fit the suitcase snugly against the toboggan's bow so she could lean back on it. Nurse Brawn wrapped the patient in blankets. Underneath, between Hazel's feet, Dr. Hamilton had placed her medical bag. Rodney Ross and Jim Gillies then pulled the ropes attached to the sled and it moved forward atop the snow, the others following. The two loyal dogs, tails still wagging their excitement, trailed again in the snowshoe tracks.

Phil and his son stood on the front porch. The Tardiff daughters stayed inside at the living room window, watching as the toboggan was pulled across the lawn and down onto the roadway.

"Is everything going to be all right?" Mary Lou asked Phyllis, who was now the woman of the house, having promised her mother.

"Of course, it is," said Phyllis. She reached for Mary Lou's hand and held it. But the younger girl could hear the worry in her sister's voice.

Sitting as Hazel was on the sled, they could see their mother's smiling face. Following behind the toboggan, the sides of her long coat flapping in the wind, was Dr. Hamilton. The family watched until Virginia Hamilton's blond hair and brown coat disappeared around a turn in the road. It was over two miles back to where they had left the borrowed Chevy.

JIMMY HAIGH COMES HOME

It was still early afternoon when the shiny black station wagon pulled into the driveway of Wiggin Funeral Home in New Hampshire. It circled around the building and backed slowly up to the door used for loading and unloading. As the driver opened the rear doors of the car, two workers came out to assist. They slid the litter out of the back, Jimmy Haigh's body inside the bag that rested on top. There was no room and no need to drive a casket up to Maine. That was not the procedure unless the burial would take place there.

Now Mr. Haigh could be dressed in the clothing that had arrived for him that morning. He and Harland Davis had already been embalmed in Thomaston. The funeral home there had billed $25 for the embalming, and $15 for the removal of the body from the wharf at Port Clyde. Out of courtesy to the Haigh family, they would not charge for the two nights the body had remained at Thomaston Funeral Home. In turn, Wiggin would add the Maine expenses to the Haigh family bill. The trip up to Thomaston cost $65. Another $10 would be charged for opening the ground for burial in the spring, after the last frost. That's when they would remove the Haigh baby, the stillborn boy, and place his box on top of his father. It's what Mrs. Haigh had requested. In all, including the mauve casket and funeral service, the bill would total $730.

BATH MEMORIAL HOSPITAL

It took over an hour to get Hazel to the waiting car on Lover's Retreat Road. From there, it was a matter of minutes to the hospital in downtown Bath. She was checked into a room with two other expectant mothers and made comfortable. After another quick examination by Virginia Hamilton, it was determined the baby would make its entry into the world within a few hours. Hazel asked the doctor to telephone Phil and the family to let them know she was safe in a hospital bed. Then she lay back on her pillow to wait, the newest issue of *Reader's Digest* to keep her company.

UNDER THE SNOW

Seaman 1st Class Paul Delaney was seated at his mother's dinner table. In the center was the pilgrim salt and pepper shakers, the pilgrim man holding the pepper, the woman the salt. Somehow, it seemed right. Between the shakers sat the ornamental turkey, its crepe-paper tail fanning out proudly. The whole house smelled of cooking, the roasted bird, the stuffing, the smell of pumpkin pies. After the blessing, his father passed him the bowl of mashed potatoes. His little sister put the sauceboat filled with thick gravy near his plate. Paul had never been so hungry as he dug into the carrots and stuffing.

He woke to the rumble of his belly, growling from the hunger he had been feeling for hours. His last visit home had been before coming to the naval base in Maine, when he arrived in Staten Island for Thanksgiving with his family. Even the smells of the cooked foods had been real to him. But now he was faced with a dark, cold reality. He'd been dreaming.

He sat up and rubbed his eyes. Why hadn't he bought a few chocolate bars for the trip, as sailors were advised to do during winters stationed in Maine? He had drunk his second orange pop the day before. Several times he had opened the side glass as much as it would go and packed the empty bottle with snow. Put under his coat, next to the warmth of his body, the snow had eventually melted. With so much snow around him, he would always have something to drink. But he was hungry. His watch said five o'clock. Was it evening? If it was, he had now been trapped beneath the snow for almost two days.

JIMMY HAIGH'S WAKE

Barbara Ann watched from the top of the stairs as they carried her father's casket into the house and arranged it in front of the three windows that formed a bay in the parlor. Her mother had already reminded Bubbles how Jimmy loved sitting by those windows on Sunday mornings with his newspaper, the sun casting shadows on the rug at his feet. She went back into her bedroom, Skybow at her heels. She sat on the end of the bed, tears in her eyes, and waited. She heard voices saying goodbye, finally, and then the front door closing.

Bubbles knocked on the door. She came into the room and sat on the bed next to Barbie.

"Your mom wants a few minutes alone with your dad," Bubbles said. She pushed Barbie's bangs back from her eyes and then hugged her. Barbie hadn't wanted anyone to see her tears, but now they filled her eyes. She had promised herself to be brave for her mother. It's what her father would want. She recited again the words she had adopted as her motto when she first heard of her father's death. *This is how it's going to be from now on. This is your new life, Barbara Ann. So be a big girl and don't cry.*

When her mother called for her to come downstairs, Barbie stood and smoothed her dress. Bubbles took a hairbrush from the dresser and brushed her hair.

"There," she said. "You look beautiful."

Down in the parlor, Ellie beckoned to Barbie to come stand with her at the casket. Her father looked handsome, his head resting perfectly on a light gray pillow. He wasn't wearing his glasses, so he looked like he did when he was swimming or getting ready to go to bed at night. There was a peaceful look on his face and it comforted his daughter to see it.

"So handsome," Ellie said now. To Bubbles she said, "His wrist is broken." She pushed the cuff of his right jacket sleeve up to show them the discolored swelling. "He must have done his best to come home to us."

She leaned forward then and kissed her husband's face. Barbie reached to touch the folded hands.

"Daddy," she said.

"Now he's home," said Ellie. She brushed the top of Jimmy's hair with her fingers, something she often did when he was sitting at the table waiting to eat. "Go put on your best dress, sweetheart. Your grandparents will be here soon."

When Barbie left and went back up to her bedroom, Ellie looked at Bubbles.

"This will be the hardest thing I'll do in my life," she said. "I thought it would be losing the baby. But I was wrong."

Bubbles embraced her friend. She wished she could do more.

"I'll go put the food out," she said.

THE ASPEY HOME IN WARREN

Dr. Charles North waited for the road to Warren to be well plowed before he drove the sky-blue DeSoto up to the Simmons Funeral Home, ten miles away. He noticed smoke belching from the tall chimney of the Georges River Woolen Mill as he drove into town. With Chester Wallace working through the night in his army-truck-turned-plow, the roads had been cleared and the mill was back up and running.

This task would not take long. There was no doubt how George Aspey had died, sitting in his chair after having walked a good distance in heavy snow. Knowing the circumstances of the death, knowing there was no foul play involved, a cursory examination was all that was needed. Charles North wrote on the death certificate that George had struggled in deep snow before *succumbing to a heart attack due to coronary occlusion*. He signed his name at the bottom and then dated the certificate.

Meanwhile, over on Ridgemont Drive, Hilda was doing her best to keep her mother from weeping. A few friends had gathered for coffee and small talk about George's life. As a Christian Scientist, Annie did not want a major fuss made over her son, and knew he would not want one, either. George's older brother, Harry, was hoping travel was possible for him to drive up from Boston for the funeral on Thursday. They would handle details as they were faced with them. One decision had been made that gave Annie some peace. With George gone, her only daughter would move back home and into his bedroom.

BREWER

Since the small city of Brewer was farther inland and not as affected by the amount of snowfall as other parts of the state, the A. J. Tucker Shoe Factory had opened for business as usual. Pete Godley went back to his managerial job with plans to invite pretty Margaret Hatch to a movie that coming weekend.

The Greatest Show on Earth was playing, and it was loaded with big-name actors. He didn't even care that the Morrill brothers had begun teasing him about puppy love and stars blinding his eyes. Pete had a feeling about this girl. He figured she was a keeper.

Jimmy had stacked boxes in a back room of the tavern most of that day. His plan had been to clean the room when warmer weather arrived, and all the doors could be thrown open to welcome in spring. Now he would save that day for fishing instead, maybe even on Branch Lake. His and cousin Pete's adventure tale was still being recounted, especially at the bar. But the day was coming when it would grow old and new adventures would need to replace it.

There could have been a different, even tragic outcome. But they had done the right thing by staying put in a blizzard. If there was one thing the military had taught both men it was that human beings are expendable as far as nature and luck are concerned. Not tempting either one was advice to live by.

STICKNEY HILL

The Pomelows were a family that had already seen hard times. Sonny's sister knew that her father adored her, his only daughter. But Ray Sr. had that quick temper. Once, when his wife complained of having to wash dishes, Ray grabbed up the four corners of the table cloth and tossed the dirty dishes out the back door. "There," he said to Grace. "Now you won't have to." But there were moments of dignity. Louise graduated from high school by holding down jobs and caring for young Sonny. And there was compassion. The fancy shoes that Louise treasured were gone one day when she got home from school. Grace had given them to a girl who had far less possessions than the Pomelows.

When the sad news reached Louise Pomelow Joslyn that her little brother had been killed, she was devastated. It was the last kind of news she imagined ever receiving. Her mother, her father, other grandparents might pass away and she would be surprised and would grieve. But young Sonny? Just fifteen years old? It was a hard Monday night she spent down in Etna, the roads still closed so that she couldn't drive home to comfort her parents. Now she had arrived that afternoon with a small suitcase in hand to do what she could to help.

Grace was lying in bed, unable in her grief to lift her head from the pillow. Other family members and friends had arrived to begin food preparations for the wake that following night. The road conditions were delaying all manner of events, from weddings to funerals. Grace and Ray Sr. would wake their only son at home. The front room would be cleared to make space for his casket. And they would set up folding chairs borrowed from the elementary school next door.

The first thing Louise did when she arrived at the house was to comfort her mother. She put an extra quilt over Grace, who had begun weeping again. Then Louise went into her little brother's bedroom and sat on his bed. Ten years younger, he had been more like her son. Sonny had taped pictures of cars to his walls, all pages torn from old magazines. A pair of black PF Flyers, sneakers he was proud to own, lay on the floor near a pair of socks. He likely wore his winter boots when he hitchhiked to the Y the day before. Louise touched the Boy Scout medals proudly displayed on Sonny's dresser. She would be sure to arrange them on his casket when he came home the next day.

With friends and family members now gone, twilight moved in over Stickney Hill. Louise put on her coat and gloves and stepped out into the fresh air. The sky was bluish-gray, the horizon still lit with refracted remnants of daylight. The storm was over. It was gone without a thought of what it had left in its wake. The elementary school sat next door, its black windows silent. It was there that so many memories had been formed over the years, for both her and Sonny. She had been in the last class to graduate high school there. And Sonny had attended classes up to the eighth grade when it became an elementary school.

Overhead, the largest stars began to appear as soft sparkles. The sound of the eight o'clock B&A train would soon echo up from the cove on the Pleasant River. The trains were running again. That lonesome melody of the whistle, Louise knew, would follow her for the rest of her life, a sad reminder of the boy she had hoped to protect forever. And how could she drive over that railroad crossing again without remembering?

Seeing her father moving about in the living room, his shoulders stooped with the sadness they now carried, Sonny's sister said goodnight to the stars.

DREAMING A LOVE STORY

Paul Delaney had stopped checking his watch to the glare of his cigarette lighter. It didn't matter anyway. It wasn't like he was going somewhere. If anyone turned up, he would be still buried there in snow, regardless of the time. He dozed off again, finding it difficult now to stay awake. He had worried earlier when he relieved himself again through the opening in the backseat door that he wasn't getting enough oxygen. Maybe that was why he so was sleepy. He longed for a cigarette, but had managed to refrain. He needed his lungs working, and he needed all the air inside the car. Was he running out of the supply? Is that why he wanted to sleep all the time now? So be it. It would help the time pass.

There was no way for Paul Delaney to know that the original idea for the naval base where he was stationed at Winter Harbor had begun in the mind of Alessandro Fabbri back at the turn of the century, and not far from where the young seaman was now buried in snow. When Great Britain had declared war on Germany, on August 4, 1914, Bar Harbor had been aflutter with excitement. The magnificently outfitted German ocean liner SS *Kronprinzessin Cecilie*, loaded with $11 million in gold and silver and considered a prize in wartime, had been mid-ocean on its way from New York when news of the declaration arrived. The captain, fearing capture by British warships, quickly turned his ship around and headed back to neutral American waters. For days rumors flew around the country and world. Where was the ship? Captured by the English? Safe at a German port? It had actually dropped anchor in Frenchman's Bay, several miles from the rock promontory at Otter Cliffs where Alessandro Fabbri experimented with his wireless telegraph, and a couple miles from where Paul's car was covered in snow.

The war was brand-new, just two days old, and so the Fabbri brothers and residents of Bar Harbor welcomed the German captain, crew, and the nearly three hundred first class passengers with generous entertainment that included banquet dinners and the finest wines. Twice a week, Mrs. Fabbri bought all 620 seats at the newly built Star Theater, which stood across the street from the Criterion where Paul had taken Mona to the matinee on Sunday. Even the lowly passengers in steerage could watch a silent film reel of *The Perils of Pauline*, the first installment having been released just that April.

The usual wartime rumors were soon circulating. Fears and suspicions ran high, and the enemy was believed hiding in every shadow. And what about those brothers with foreign-sounding names? Just the notion of *wireless* communications seemed ominous enough. With gossip growing, it kept leading back to the beautiful summer mansion named Buonriposo, and those high and unimpeded cliffs at Bar Harbor. GERMAN WIRELESS AT MOUNT DESERT? was soon a local headline question.

By autumn of 1914, even the *New York Times* pointed a finger at the affluent Fabbri brothers of Bar Harbor. Hadn't they lavishly entertained the German passengers when the ship anchored there? America was still not at war, but if an American was coding vital information and transmitting it across the ocean to Berlin, it would violate America's neutral status. And while the German "treasure ship" was anchored in the waters at Bar Harbor, wasn't it true that Alessandro and Ernesto, instead of enjoying the fine banquets and *The Perils of Pauline*, took turns locked in their "wireless room" for days at a time to transmit messages? No, it was far from the truth, but in that charged atmosphere the truth didn't matter.

At first Ernesto Fabbri assumed it a joke that he and Alessandro were thought to be spies in league with the Germans. "I guess we were too kind to the poor fellows," he said of playing host to the hapless crew of the *Kronprinzessin Cecilie*. The wireless setup on Mount Desert, he assured journalists, was the result of his brother's passionate hobby. When the United States entered the war, it was Alessandro's chance to prove loyalty to the country of his birth. He offered his superb wireless telegraph station to the United States government as part of the war effort. Three years later, at Ernesto's and Edith's mansion in New York City, Alessandro fell ill suddenly and died. He was buried in the Vanderbilt Cemetery and Mausoleum. Edith, having fallen in love with Alessandro by this time, was heartbroken. As if to profess her undying love, she divorced Ernesto less than a year later.

But the dream Alessandro left behind eventually blossomed, thanks to John Rockefeller's enormous wealth. It was now alive in the naval base at Winter Harbor. When Paul Delaney drove past Buonriposo minutes before skidding down Ireson Hill, he was driving past not just a forgotten love story, but a part of naval history. It was thanks to that history he was now stationed in Maine.

A NEW LIFE

The city of Bath was recovering from its worst northeaster in many years. The plows had been towed out of snowbanks and pulled from drifts and were back at work clearing the streets. When that was done, they would concentrate on the rural highways leading into town. The decision made to bring Hazel Tardiff to the hospital was a wise one. She had been in active labor since six that evening, and the baby was due at any time. With the streets still being plowed, Virginia Hamilton had decided to spend the night at the hospital. She was exhausted from snowshoeing four miles that day. She had changed into her white slip and a pajama top, and was napping in one of the beds reserved for doctors and nurses when she was awakened. Hazel was ready to deliver her baby.

It was just past nine o'clock that Dennis George Tardiff was born, his cry breaking the stillness of the hospital's small maternity ward. A healthy baby, he weighed nine pounds and ten ounces. The *George* was for Hazel's father, that fisherman from Isle au Haut who once had a nickel left in his pocket after feeding his island family over the long winter. The nurse promised to telephone Phil with the good news once she got back to her station. Hazel was thinking of how excited the girls would be, and how they would help care for the new baby in days to come. David was now a big brother. She wished she could talk to the children just then. Seeing their pregnant mother pulled away on a toboggan must have been unsettling.

Hazel dozed off just before midnight. It had been a busy day.

55 GATES STREET, PORTSMOUTH

Barbara Ann Haigh had done what she promised. She was brave in front of her grandmother and did not cry. She sat up straight in her best dress as Mrs. Haigh stared straight ahead at her son's pallid face. Friends and family members filed in and out of the parlor, each stopping to pay their respects to Jimmy. Bubbles kept fresh coffee perking and filled the platters on the kitchen table with more sandwiches and pastries once they emptied. Slowly, one by one, the mourners said their goodbyes and put on their coats. When Jimmy's parents left, Ellie stood with her daughter in the doorway and watched the car's taillights disappear at the end of the street.

There would be one more night for the wake, with new faces coming to say goodbye to a good man. With the visitors gone, Ellie went to the kitchen for a sandwich. She had not eaten all day. In the parlor, Bubbles closed the lid of the casket. She planned to stay a couple more days to help out. She went upstairs to change into pajamas. Barbie took Skybow out for his last pee of the evening. When she came back inside, she left his leash hanging on the coatrack by the door.

"Goodnight, Mama," she called to Ellie.

Ellie had just put out the kitchen light. She came to hug her daughter and kiss the top of her head. She watched as the little girl climbed the stairs, Skybow following. Ellie had already announced that she intended to sleep on the parlor sofa both nights that Jimmy was being waked. She wanted to be close to her husband. She went to her own bedroom to change into night-clothes and bring a pillow and blankets to the parlor.

In the upstairs bathroom, Bubbles was telling Barbara Ann about the dog she had owned as a girl while Barbie brushed her teeth.

"Bubbles!" they heard Ellie shout from below.

Ellie was standing at the venetian blinds in her bedroom window, wearing her pajama bottoms and still in her bra. She was staring out into the night.

"What is it?" Bubbles asked.

"Someone was standing outside my window." Ellie was shaken. "A man was out there watching me."

"Mama!" Barbie said, suddenly frightened. Ellie pulled her daughter close.

"It's all right, honey," she said. "We're safe in here."

Bubbles was the one who telephoned the police. After they arrived, she took Barbara Ann upstairs to bed.

Now fully dressed, Ellie waited in her bedroom as she peered again through the venetian blinds. In her backyard, beams from police flashlights searched the deep snow as two officers investigated. She saw one kneel not far from her window and brush away snow with his glove. She assumed he found something dropped there. Ellie felt as though she might be sick to her stomach. This was an added stress she didn't need. When the men knocked on her front door, she let them inside.

"There are reports of a Peeping Tom in this area," one officer said. "We found tracks in the snow next to your bedroom window. And we found these."

He opened his gloved hand to show her three cigarette butts. Ellie tried to remain calm. It seemed like a dream, or a scene from a scary movie.

"He likely read in the papers about your recent loss, Mrs. Haigh. Guys like that are pretty sick."

"What should I do?" she asked.

"Keep your doors locked and window blinds closed at night," the officer told her. "We'll be watching this street."

Despite being exhausted, Ellie lay awake on the parlor sofa until the sky outside the bay windows grew lighter. It was difficult to sleep knowing that a disturbed stranger had been lurking outside their home. What if he found a way inside during the night and did harm to Barbie? She slipped from under the blankets and went to Jimmy's casket. She lay her hand on the closed lid. She would lift it later, before the stream of visitors began again. For now, she wanted to say a prayer just for her husband. And then one for herself.

"Keep him safe, Lord, until we join him," Ellie whispered. "And give me strength."

Then she drew back the bay window curtains Jimmy had installed for her just months earlier. Pink veins laced the sky. It looked like the sun might break through.

BILL DWYER'S HOUSE

Bill had tired of asking neighbors up and down his street, and on streets nearby, if they had seen a yellow cat. While Snooky had never been away this long before—it was now three days—Bill was still hopeful. Cats sometimes make plans without letting an owner know. It's just that Snooky hadn't been one of those cats. He loved his regular meals and snoozing on the sofa too much to be delinquent.

Earlier, Bill had cleaned snow away that had blown into the box he left on the porch. He put a fresh bowl of tuna fish in one corner. When Snooky finally did mend his errant ways and come home, he would likely be hungry.

When Bill rose in the middle of the night to relieve himself, there was no sign of a yellow cat.

PART FIVE

PART FIVE

FEBRUARY 20, 21
WEDNESDAY AND THURSDAY

Even as our cloudy fancies take
Suddenly shape in some divine expression,
Even as the troubled heart doth make
In the white countenance confession,
The troubled sky reveals
The grief it feels.

This is the poem of the air,
Slowly in silent syllables recorded;
This is the secret of despair,
Long in its cloudy bosom hoarded,
Now whispered and revealed
To wood and field.

—"Snow-flakes" by Henry Wadsworth Longfellow

A NORTHEASTER BABY

Hazel Coombs Tardiff slept soundly through the night. She woke as a nurse placed her newborn son in her arms. She fed the baby his bottle as her own breakfast was wheeled into the room. There was a sense of

elation that it was over and had gone well, if not according to plan. The
worry of the past few days was lifted from her. She had not voiced to Phil,
and certainly not the children, that she feared having the baby at home. She
had never forgotten the story of her father carrying the small coffin down
the road to Coombs Cemetery, back on Isle au Haut, as her mother wept
in the bedroom.

The morning paper lay next to her coffee and toast. Babies had
been born all over New England during the storm. She read about the
expectant mother of eight children, being pulled by sled from her isolated Maine farmhouse to reach the home of a doctor. Hazel smiled
to read that Mrs. Arthur Smith of Mount Vernon had had a healthy
baby girl. The paper was filled with mentions of Bernard Webber. Hazel
knew his mother Annie Knight of Bath, and his grandparents. Bernie
and his crewmen were now national heroes for rescuing men from the
broken stern of the SS *Pendleton*.

Another story related to the sinking ship caught her eye. Pictured was a
young man from Portland named Carol Kilgore, one of the fortunate survivors who made it into Webber's rescue boat. Kilgore, a pantryman in the
ship's kitchen, had run away from a troubled home just the month before.
He told a friend he planned to join the Merchant Marine. His father and
stepmother assumed he was on the SS *Pendleton* since it had been docked in
the harbor at Portland the day Carol disappeared. And then Kilgore turned
up in newspaper photos, looking every bit of sixteen years old. He had been
asleep when he heard the tremendous boom that brought him racing up
to the deck. "I couldn't believe my eyes," the soft-spoken boy was quoted
as saying. "The bow was gone. We didn't think we'd ever get off the ship.
There was nothing we could do but pray. It was my first trip to sea, but I'll
go back again."

Putting on slippers, Hazel carried her baby down the hallway to the nursery.
He was sleeping soundly. A nurse took him and lay him back in his bassinet.
With Bath Iron Works now up and running again, Phil would visit later that
evening to hold his new son for the first time.

RESURRECTION

Paul Delaney was now accustomed to finding solace in sleep. He was dreaming again, this time of the thunderstorm. He had been no more than ten years old and was fascinated with the lighting cracks over his head and the booms of thunder. Standing on the street in front of his childhood home, he watched as the storm moved in. When an electrical wire snapped and fell near his feet, Paul reached down and picked it up. Rain and wind washed over him. That's when he heard his father yelling. "Put it down, Paul! Put it down and get away from there!"

His family told that story often at the supper table, about the day Paul almost electrocuted himself. When the *tapping* noise broke through his dream, he saw the end of the electrical wire as it snapped against the tar of the road. He worked toward awareness, finally opening his eyes. Had he heard it? Was it his imagination? Now there were muffled noises from above, distant, but maybe the sound of human voices. He hoped it wasn't angels he was hearing. He sat up in the back seat and crawled quickly to the front. He wound down his driver's window, flicked on his lighter, and stared at the packed snow outside. There were the indentations where he had pulled in handfuls to pack into his pop bottle. And then there it was again. *Tap. Tap.* On the roof of the car.

"Hey!" Paul shouted out his window, hoping his voice could be heard. "Hey, I'm down here!"

Maybe someone was up there with a pole, pushed down through the snow. But there was no response. And now it seemed the muffled sounds had gone away. The angels had flown back to heaven without him.

"Help!" Paul yelled. He leaned on the horn and was relieved it still worked. And then he saw the wooden pole cut through the snow just outside his window. He reached for it, but it was gone too quickly. He waited, cigarette lighter in his raised hand in case it came back. There it was again! He grabbed fast and this time caught the pole in his fist. He held tight as he pulled it, letting whoever was up there know he was there. Now the voices grew louder overhead, in an excited timbre he understood as recognition he'd been found.

He wiped tears that appeared, unwanted, in his eyes. He needed to show strength now. Things were going to be all right. He was a sailor, not a crybaby.

On a quick decision, Paul reached for one of his Chesterfields and lit it up. Never had a cigarette tasted so good. He sat back behind the steering wheel to wait for the first daylight he'd seen in three days. His whole body felt weak, cramped from no exercise, his legs in need of stretching. He would not know until later that standing on packed snow almost a dozen feet above the car were Howard "Packie" McFarland, Bar Harbor's chief of police, his son John, and a local officer. They had come to mark the depressed spot where it appeared an automobile was buried in drifts so that the plow wouldn't hit it. Chief McFarland was shocked to feel the weight of a hand grabbing onto the end of the pole he had poked down through the snow.

"We got someone buried here!" McFarland shouted. City workmen who had been clearing the road came quickly with their shovels to help.

Smoke curled from Paul's nostrils as he sat listening to the scraping sounds over his head. Twenty minutes later light broke through the darkness, causing him to blink. Gloved hands wiped snow from the windshield and Paul saw two booted feet standing on the hood. His first thought was for Jeffrey's car. Poor guy. He must have been wondering where his roommate and his vehicle had disappeared.

"Who's there?" he heard a deep voice ask. It came from a man who was now standing on the sedan's roof.

"Seaman First Class Delaney!" Paul shouted out the window.

Now shovels ate the snow from around the top of the car, creating enough space near the driver's door for Paul to get out. He still didn't realize that he'd been buried beneath more than ten feet of drifted snow.

"Can you crawl out the window?" the voice asked, and now Paul saw faces on the hood, peering in at him.

"I'll try," he told the men. He stubbed his cigarette in the ashtray and edged himself out the driver's window. Hands reached down and grabbed him, pulled him up.

"Are you all right, son?" McFarland asked, brushing snow from Paul's coat.

"I think so," Paul said. "What day is this?"

"It's Wednesday morning," McFarland told him. He had been buried in the car since early Sunday evening. Sixty hours.

"That's my friend's car," Paul said then.

"We're taking you to the hospital," said McFarland. "The car can wait."

The first thing Paul Delaney did, after a doctor pronounced him dehydrated but perfectly fine, was to telephone his parents on Staten Island.

DANIEL SPEED

The young husband and father had not regained consciousness. Two days after the wreck he was still in the hospital and on the critical list. Having sustained a serious head injury his vital signs were unstable. His wife moved in with her sister in Bangor so that she would be close to the hospital. With a babysitter now for the boys, Frances walked the State Street hill each day to visit Danny. While he lay in his hospital bed, she stayed close by, talking to him, reminding him of what he was giving up if he left them. The doctors were predicting he would remain that way for another week, perhaps longer.

Word of the wreck had been the talk of Brownville and Brownville Junction ever since it happened. And now the newspapers let the rest of Maine learn about the tragedy. A reporter had written up the story and it appeared in the *Bangor Daily News*. A photograph showed Speed's stretcher surrounded by a half dozen men who had lifted it down from the B&A passenger train. His head was wrapped in bandages. Before the ambulance could rush the patient to Eastern Maine General, a caring police officer was covering Daniel with an extra blanket.

There was no one to blame, but blame might have helped ease the grief. Should Daniel Speed have made certain of the situation on Monday afternoon before crossing the track? Almost everyone in Brownville and Brownville Junction could have made the same decision. There were whiteout conditions and the plow was not a scheduled train. Albert St. Louis, the mail carrier who crossed the tracks in front of Daniel Speed, was just seconds ahead of his own possible demise.

It was a horrible mistake, and mistakes are made in the midst of storms. Even experienced men like Harland Davis and Jimmy Haigh had learned this lesson too late. If Daniel Speed regained consciousness, he would be a wiser man.

SONNY COMES HOME TO STICKNEY HILL

Ray Pomelow Jr. had been made ready for the wake to be held at his home. He was dressed in his good suit, the one he had worn three weeks earlier as a pallbearer at his grandfather's funeral. Louise and Grace, the sister and mother who loved him, had brought his Boy Scout medals and badges down from the bedroom dresser and displayed them on the casket. His father, Ray Sr., sat inconsolable on the sofa. Aunts and cousins were making sandwiches in the kitchen and arranging plates of cookies for the wake that evening. The funeral would be the following day, with Sonny's cousins and schoolmates as pallbearers.

By the time mourners began arriving at the Pomelow house to pay their respects to the shy boy they would miss dearly, bright stars were sparkling again over Stickney Hill. And the moon was rising over Pleasant Cove where Sonny had often skated, and over Split Rock where he and the other kids swam each summer. As usual, the curfew whistle of the B&A train was due at eight o'clock.

THE LAST CHRISTMAS

On Wednesday night, Ellie Haigh slept again in the front parlor so she could be near her husband before he was taken away. Barbara Ann came to pay respects to her father before she went up to bed. The casket would be taken away in the morning. This would be the last time she could say goodbye in private.

What Barbie had been remembering all that day was their last Christmas, less than two months earlier. Jimmy had been up to Maine for lobsters then, too. On the way back, he spotted a dense fir tree growing along Route 1, with full, perfectly shaped branches. He pulled his red truck to the side of the road and waded through snow to reach it. Using just his pocket knife, he cut at the narrow trunk until he brought the tree down. He drove it home in the back of the truck as a surprise for his daughter. He helped her position it in front of the bay windows, its generous branches and the smell of balsam filling the room. The fir was so huge that Ellie had gone to the local Grants store to buy more lights and ornaments. Then she made hot chocolate as Barbie and her father decorated the tree. It was their final Christmas as a family.

That night before the funeral, Barbie dreamed she was still so small that she held Jimmy's hand as they walked to the local movie theater. They had done that often in real life. In her dreams and memories of her father, the sun was always shining.

THE FAMILIES SAY GOODBYE

On Thursday, Elizabeth Taylor married Michael Wilding at Caxton Hall in London, a building known since World War II for artistic events and civil marriages for celebrity and high society names of the day.

That same day, funeral services were taking place in Maine for Harland Davis, Charles Voyer, George Aspey, and Ray Pomelow Jr. In New Hampshire, services would be held for James Haigh. All five men would be properly buried in the local cemetries of their hometowns later that spring, when the earth had softened from the winter freeze.

Meanwhile, winds had swept northeast from the Atlantic Ocean and formed a low-pressure area off the coast of Georgia. The pot was now simmering with possibility. Should a powerful stream of frigid air originating in the Canadian Prairies travel eastward to collide with that warm air in the Gulf Stream, an extratropical cyclone could likely result and a future northeaster born. Now weather reports were ominous. Another big storm was likely to hit New England in the new few days.

And a sixth fatality for Maine was still to be counted.

BARBIE ANN REMEMBERS

What nine-year-old Barbara Ann Haigh would remember about the day of her father's funeral was the abundance of flowers, their smell overpowering the parlor of their house on Gates Street. Extra easels were brought from Wiggin Funeral Home to hold the many wreaths and sprays. Following behind the funeral hearse to Pine Hill Cemetery was another hearse filled only with baskets and vases of flowers. One large arrangement had been phoned in from a florist shop up in coastal Maine, sent from those fishermen who had come to know James Haigh from business dealings. Its white chrysanthemums were fashioned in the shape of an anchor and trimmed with a navy-blue ribbon the color of the sea.

Barbara remembered something else about that day. As they stood ceremoniously at the gravesite, it started snowing while the sun was out. A quick shower of wet flakes, fluttering like moths and settling down on the blossoms around the coffin.

"The flowers look like they're crying," she had whispered to Ellie.

The mourners said their goodbyes and slowly filed from the cemetery, Barbie and her mother last to leave. The pallbearers, Earle Sanders being one, then lifted Jimmy's casket and slid it back into the hearse. It would rest in the vault until spring, when the baby who died at birth could be buried with his father.

That night of the funeral, exhausted from grief, Ellie Haigh had finally fallen asleep. It was past midnight that she woke to the sounds of car doors slamming from the street behind her house. Excited voices rose in the night, just beyond her window. She left her bed and peered through slats in the venetian blind. A blue light flashed atop a police car parked in her neighbor's driveway. Yellow orbs bounced in the darkness as two officers with flashlights searched the snowy yard. Suddenly, the dark shape of a man broke from behind the leafless lilac bush and fled into the darkness. Ellie pulled back from the window, her heart beating.

Minutes later a patrolman knocked on her front door.

"We caught him, Mrs. Haigh," he said. "This ain't his first rodeo, either. We figured all along it was him."

MAINE FATALITY NUMBER 6

Fatalities for each storm are somewhat unreliable, as storms themselves often are. If a man dies from shoveling his driveway on the day of the storm he is counted as a fatality. If someone dies shoveling within a day or two of the storm, as cities and towns get back to normal, they are counted as fatalities. But the man who waits for a few days after a storm to shovel out his backyard and succumbs to a heart attack is not counted, even though he is moving snow from the same storm. Maine counted six people as having died in the northeaster that lasted from February 17 to February 19. With Paul Delaney now rescued from the buried car, with Hazel Tardiff having safely delivered her baby, with Bill Dwyer feeling robust except for lamenting his missing cat, who was the sixth fatality?

It was not Daniel Speed, who would regain consciousness ten days after the wreck.

A death had happened Sunday afternoon on February 17. The northeaster was slower in reaching East Baldwin, a little town thirty miles northwest of Portland and the last exit on the Maine Turnpike. As Weston Gamage was circling the churning waters around the submerged *Sea Breeze*, searching for the bodies of two men who did not deserve to die, the elderly Charles Ward was putting on his winter coat and boots. East Baldwin had once been known for its fine orchards and a factory for drying apples. Now the fields and gnarled trees were laden with snow, and more was on the way with the winter tempest just reaching the coast of Maine.

Charles had a dozen chickens to feed in the henhouse out back before the storm intensified. By three o'clock, the new snowfall amounted to just four inches. That was easy walking for any man. But Ward had not shoveled old snow from the last storm. He had been experiencing occasional chest pains that last week and was not up to shoveling. For once, he would listen to his doctor. Now, with fresh snow falling, he followed his earlier footsteps on the path to the henhouse. In one hand, he carried a pail of water. In the other was a pail of cracked corn and oats, covered with table scraps.

When Charles opened the door of the coop, he was met with a welcoming chorus of bird chatter. He filled the tin water feeder first and hoped it wouldn't freeze too early. Then he scattered food scraps over the wooden floor, pieces of breakfast toast and donut. He dumped the corn and oats into the gray galvanized feeder. The chickens gathered around his legs, excited.

"Now mind your manners," Charles told them.

His wife Lizzie had been gone for a decade. The birds were good companions. In the summers they pecked and grazed, snatching bugs and pulling up earthworms on his front lawn while he smoked his pipe on the porch swing and admired them.

It was on his way back to the house, his boots pushing through deep snow, that pain exploded in his chest. Charles knew it was his heart. He had been taking medicine for years. The doctor warned him it was time to slow down. "You're seventy-six years old, Charlie. You're not a young man anymore. Pay a neighbor boy to split your firewood." He saved money by putting up his own wood, every dime needed. And how could raising a few chickens hurt? He always had fresh eggs. They were buying them in stores these days. It would get to a point where youngsters wouldn't know where eggs came from.

Charles dropped the two pails, soft splats in the snow. The intense ache spread across his chest and he went down on his knees. Behind him he heard the rustle and contented murmurs of his chickens. He had built that small house with his own hands twenty years earlier. It was patched up in places where the winters had been harsh to the wood, but it was a fine coop. "Make them a door for winter," Lizzie had said, watching him hammer nails. So he had cut an opening in the back and covered the outside with wire to keep away the foxes.

Memory. When he left Fredericton, New Brunswick, and followed his parents to Maine he was nineteen years old, with blue eyes and thick hair that was shiny black. He never became an American citizen. There is something about *home* that is in the blood and cannot be erased. It's what the migrating birds know each spring. A laborer all his life, Charles married for the first time at age forty-six. Lizzie was four years older and it was her second marriage. Their gift to each other was companionship.

Charles Ward fell forward into fresh snow. It filled his nostrils and mouth.

"This is not the place for a man to die," he thought. "A man should die in his bed."

How long before his neighbors wondered about him and came to check? They knew he prized his solitude and independence, he had made sure of it. Maybe he had been too standoffish at times, accepting a loaf of homemade bread or a jar of apple jelly with a curt thank you before he closed the door. In his last breaths, feeling the cold snow inside his nostrils, he thought of his chickens and how they had enough food. He could thank Lizzie for the little door. They would have snow for water until someone came.

"What's wrong with raising a few chickens?" Charlie had asked himself many times.

Over the course of that night and for the next two days, Charles Ward had been slowly covered in four feet of windblown snow. Lying six feet from his henhouse, he was discovered on Wednesday morning by a concerned neighbor who realized she had not seen the oil lamp glowing in Mr. Ward's window for the past three nights.

The local newspaper reported that Mr. Ward "succumbed to a heart attack after struggling to shelter." This information likely came from the doctor who examined the dead man and wrote the medical report. But no one will know what really happened. It's possible Ward's was a sudden death and he died right away. Or he died within a few minutes of falling, of cardiac shock, unable to get up. Or, with his heart compromised and being unable to walk to his house, he might have died several hours later from exposure, as the night grew colder and the falling snow slowly buried him.

This death would be officially recorded three days after Charles Ward died. But on Sunday afternoon, February 17, 1952, Mr. Ward became one of the six northeaster fatalities to be counted in the state of Maine.

PART SIX

THE EPILOGUE

What matter how the night behaved?
What matter how the north-wind raved?

O Time and Change!—with hair as gray
As was my sire's that winter day,
How strange it seems, with so much gone
Of life and love, to still live on!

Henceforward, listen as we will,
The voices of that hearth are still . . .

—"Snow-Bound: A Winter Idyl"
by John Greenleaf Whittier

Out of respect to Harland Davis, the fishermen on Monhegan Island refused to cash the checks he had paid them for their lobsters. One man kept his in the family Bible where it was found after his own death many years later.

Alice French Davis slowly descended into sorrow and depression after Harland's death. In her deep grief, and under pressure from her mother and the family doctor, she agreed to give her baby son up for adoption when he was born later that summer. She was reminded that she already had a young daughter to raise, with no means of income.

For the rest of her life, Alice regretted that decision, at times overwhelmed with remorse. Six years after Harland died, she would marry again and have four more sons.

That child she gave up, Harland's son, was adopted by a couple named Wilson. Bill Wilson grew up in Bath, nine miles from where his birth mother lived and raised her new family. She was Alice Wilson then, having married a man with that same surname. When Bill was five years old, his adopted parents took him to meet Alice for the first time. He remembered that a pretty girl named Carolyn went walking with him along the water. He would not see them again until he was a senior at Bowdoin College. That's when he learned Alice was his birth mother and Carolyn his half sister, the little girl who had loved Harland so. This was also the day Bill Wilson learned that he had four half brothers.

Bill and his wife, Sandy, raised their own family in Woolwich. He spent his career at Bath Iron Works, retiring after thirty-five years as a senior buyer and contracts specialist. He always felt fortunate that he met his grandfather Riley Davis before the older man's death, and was even taken by his adopted parents to Riley's funeral. He is still in close contact with his half brothers.

Over the years, Bill Wilson has collected photos and memories of the father who died before he was born. His favorite photo of Harland is one where he is kneeling on his beloved *Sea Breeze*, a lobster in his hands, proud of the new wharf he'd just finished building. That's the wharf Elisabeth Ogilvie mentioned in her 1950 book, *My World Is an Island*. Bill sees in his grandson, Kyler, some of his father's characteristics, or so he's been told by those who knew Harland. Like Harland, when Kyler sets a goal, he follows it through to the end. His favorite hobbies are repairing boats, deep-sea fishing, and pulling lobster traps for a friend. Quite often, even during the winter months, Kyler likes to swim in the cold waters of the ocean off Maine's ragged coast.

Five years after being buried alive in snow, Paul Delaney married the girl of his dreams. He and his wife, Patricia, bought a home in Staten Island that they referred to as their "forever home." They raised their family there. Paul often told his children about being trapped in a car during the northeaster of 1952. An avid golfer, he enjoyed a long career as an insurance underwriter. In the 1970s, as if in tribute to his once having been in the navy, he took up sailing. Paul passed away in December of 2004.

In 2002, the navy closed the Winter Harbor base where he was stationed. Alessandro Fabbri's dream and John Rockefeller's costly vision might have ended there. But the National Park Service has since established the Schoodic Institute for research in science and education.

◆

Daniel Speed was unresponsive for ten days. When he regained consciousness and learned that his passenger had died, he knew he would live with that knowledge for the rest of his life. As with any caring human, Speed would wrestle with guilt. Why hadn't he been more careful? Why had he followed Albert St. Louis's taillights without paying closer attention to the track? Why was he alive and young Sonny dead? All the kid wanted was a ride.

For the next six weeks, as he recovered at his mother's house, Danny ate liquid meals through a straw. He had time to rethink the steps of his life so far. Memories visited him often, usually at night. The war had ended while he was finishing basic training in Georgia. He went by ship to France, and then traveled on into Germany by railway. At Landsberg, his was assigned duty as a prison guard. The prison there had been taken over by the Allies in occupied Germany to house Nazi war criminals. The place was already famous. During his short imprisonment in 1924, this was where Adolf Hitler had dictated *Mein Kampf* to Rudolf Hess. When Danny arrived it was home to one of the war's most famous criminals, Dr. Claus Schilling, who had conducted malaria experiments at Dachau on over a thousand Jewish victims. The doctor was hanged in May of 1946, with military personnel on hand to film the incident.

When Speed was discharged in October, he left Germany and went down through Austria and Switzerland, into Italy. He signed his name in the guest book at the Roman Colosseum before coming home to Maine and doing hard labor on his father's farm. Two years later, he saw Frances Harris for the first time. Her nickname was Mike and she was waiting for the school bus. Three weeks later, he asked her to marry him. Her brother drove them to East Bradford to get a marriage license from the town clerk. It was a cold and snowy day and the truck skidded on icy roads. They ended wheels-up in a brook, but all three managed to climb out unhurt. The brother got a ride home, but Dan and Frances walked the rest of the way. They got the license and were married that December.

In a world that was sometimes on the brink of destruction and madness, Frances represented all that was good. She had been the best decision in his life. She would never forget that day when a constable knocked on the door of their upstairs apartment to tell her Danny had been in a terrible wreck and was unconscious. His passenger, Sonny Pomelow, the grandson of her landlord who lived downstairs, was dead.

When Danny and Frances moved into his mother's house where he could convalesce, the minister visited from the Bradford Baptist Church. Danny and his young wife gave their lives to the Lord that night. They brought him into their hearts and never looked back.

The Speeds raised four sons and a daughter. When Daniel died in 2008, his obituary read like a list of occupations and hobbies from which a single man might choose a few: farmer, logger, railroad fireman, carpenter, builder, home inspector, church deacon, softball umpire, beekeeper, gardener—he was known for raising canna lilies—pool player, hunter, fisherman, husband, and father. Was he just ambitious, or was he doing all he could with the life he believed the Lord spared that day at Brownville Junction's second railroad crossing?

It's likely Daniel Speed never knew this while he was at Landsberg Prison. A German horticulturalist had founded a nursery at Stuttgart, ninety miles away that was, and is to this day, instrumental in developing many flowers, including the beautiful canna lily, which was Danny's favorite.

◆

Dr. Virginia Clay Hamilton was a woman ahead of the times. Coming from what would proudly be referred to as "an old Kentucky family," Virginia spent her summers at Hillcrest, the family's country home near Lexington. The rest of the year she was at their house in Louisville. Her father was Arch Hamilton, a state senator who drafted the first voting-machine bill while in office and was then instrumental in getting the machines adopted by Kentucky officials in 1942. He was a fierce advocate for education, especially for the underprivileged. Arch had supported and worked hard for the new concept of the Moonlight School in Eastern Kentucky, a night-school program to teach reading to adults.

Surrounded by this atmosphere of public service and the importance of education, Virginia was more scholar than debutante. In 1917, as World War I was still burning up Europe, she wrote her thoughts down in a poem, her hope for mankind. She was fifteen years old and "The Happiest Hour" appeared the following year in the popular magazine *St. Nicholas*, having won first place. In writing about primal man and her wish for the future, the teenaged Virginia wrote: "*We think we have left him far behind / We've abandoned club and spear / Yet we still war on with brain and brawn / Moved mostly by hate and fear.*"

Virginia and her husband Boyd Bailey raised a son and a daughter. She became a bit of a local heroine when the newspaper ran photos of her on snowshoes, following behind her pregnant patient being pulled on a toboggan. A decade after that storm, Dr. Hamilton retired from delivering babies in Bath. She and Boyd moved twenty-five miles south, to the shores of Casco Bay. She died there in 1981, at the age of seventy-nine. During the writing of this book, the curator at the Henry Clay Estate in Kentucky was sent the article from the *Bath Independent* about the desk that Dr. Virginia Clay Hamilton, formerly of Lexington, had in her home office on Washington Street. She had been given it when she was fourteen years old by one of Clay's descendants, or so the story read. The curator could not verify that Henry Clay, the "Great Compromiser," had ever owned such a desk.

◆

The death of such a young boy stunned the community. The entire freshman class attended the funeral for Ray Pomelow Jr. So did Troop 111 of the Boy Scouts, wearing their full Class A uniforms with long pants. Come spring, Sonny was buried in Pine Tree Cemetery in Brownville Junction. When his class graduated in 1955, one page of the yearbook was dedicated to his memory, with particular mention of his "quiet personality and courteous ways."

In the senior Ray's last years, he lay bedridden with arthritis in the front room where his only son had been waked. The pain of loss grew into their lives, a silent but ever-present visitor. Yet no one wanted to talk about the day of the train wreck. Ray told his grandchildren entertaining stories instead, and they remember him that way. But none of them recollect their Uncle Sonny. Like a shooting star, the boy disappeared too soon to be known by the next generation.

Before she passed away in 1996 at a local nursing home, sister Louise left an important story for her daughter to safeguard. The night before Sonny Pomelow died, Louise sat up in the darkness and watched as a white light moved slowly around the circumference of her bedroom ceiling, touching each corner before it disappeared. She was too scared to wake her sleeping husband. But for her entire life Louise would remember her thoughts at that moment in time. "Oh, something is *very* wrong." Only the next day would she hear the news: One of the brightest lights in the heavens had fallen from the sky over Stickney Hill.

When Sonny's best friends, John Ekholm and Bob Williams, graduated from high school without him, they pooled their money as the three boys had once planned, sitting in the Naugahyde booths at the Rexall drugstore. John and Bob headed to California in John's 1947 Studebaker. Now and then as they rolled west through cornfields, or met thunderstorms head on, or felt the desert heat rolling up through the floorboards, one would look at the other and say, "You know what? I think Sonny is riding with us."

These days, when he feels the need to remember, John Ekholm grabs a beer and drives over to Pine Tree Cemetery. He parks his truck on the roadside near the front gate. Sonny lies buried next to his parents, Ray and Grace. Other Pomelows are nearby, including his grandfather and sister Louise. John stands there at the foot of Sonny's grave and drinks a beer while he pays respect to the friend he lost in the northeaster of '52.

◆

Six weeks after spying on a widow's grief, the man responsible for lurking beneath Ellen Haigh's bedroom window was sentenced to six months in jail. It was not his first arrest for this obsessive behavior, nor would it be his last.

Needing to support her daughter, Ellie took over the James B. Haigh Lobster Company. The loyal Earle Sanders stayed on, still delivering lobster and seafood to valued customers, sometimes on trips up to Maine. They continued what Jimmy had started, storing their lobsters in crates in the back channel of the nearby Piscataqua. Ellie was doing well enough that she leased a dock platform for loading down by the water. Jimmy would have been proud. But Ellie Haigh fell ill, unexpectedly, and was rushed to the New England Deaconess Hospital in Boston. She died of an aneurism shortly after being admitted. She was thirty-four years old. It had been less than two years since her husband Jimmy died in the freezing waters off coastal Maine.

Now an orphan, Barbara Ann said goodbye to her dog Skybow, who was taken in by Earle Sanders and his family. Just eleven years old, she went to live with a paternal aunt in Massachusetts. After three years of unhappiness, Barbara wrote asking her Grandmother Haigh for help. She missed the ocean. But more than anything, she missed Skybow. That's when Bubbles Finnigan and her husband offered to adopt her. The teenager adjusted and thrived. When Bubbles died a year later, Barbara Haigh remained with Mr. Finnigan as part of his family.

On New Year's Day in 1954, Earle Sanders bought the James B. Haigh Company from the Haigh estate executors, renaming it the Sanders Lobster Company. He figured his former boss would like the idea. When he retired in 1986, his son Jimmy took over the business. Earle died in 2008. Sanders Lobster Company is still doing a successful business to this day.

Barbara Haigh graduated from Portsmouth High School and then a nursing college in Massachusetts. She joined the U.S. Navy Nurse Corps. After five years, she left as a full lieutenant to have the first of her four children. During the writing of this book, she was reconnected with the Sanders family and spoke by phone with Mrs. Phyllis Sanders, who had just turned ninety-five years old. Barbara was always grateful that the family took in Skybow. "He was very happy with them," she remembered. "I would see him once a year when my aunt took me to New Hampshire for a visit."

◆

Peter Godley married Margaret Hatch and remained in Brewer for the rest of his life. He and Margaret had two children, a daughter and a son.

In 1969, during his retirement, Peter sent a letter to the editor of the *Bangor Daily News*. He was replying to a complaint from a previous letter-writer that a guided missile destroyer built at Bath Iron Works for the West German navy would be christened *Rommel*. The Field Marshall's widow would be there for the ceremony before the ship slid down into the waters of the Kennebec River at Bath. The writer had bad memories of being a young soldier in Africa when Rommel was on everyone's mind, and Germany was the enemy. Peter felt he had his own memories of that time and place. And he still carried a healthy respect for the Desert Fox, who had once praised British soldiers for their valiant fighting.

> *. . . during World War II, I was a corporal in the British Eighth Army and feel I knew Rommel well. I slept for two and a half years with nothing between me and the desert sky. The nights were cold but the days were hot and we suffered from flies, desert sores, dust-filled pores and red-rimmed eyes . . . we later realized this had all been fun until Rommel arrived on the scene.*
>
> *He made us really miserable the way he kept us "flapping" and running in circles. Desert warfare is something like being at sea*

*because we occasionally recaptured some of our comrades who had
been taken prisoner by the Germans a day earlier. Some spoke of
shaking hands with Rommel who asked if they had any complaints
about the way the Germans were fighting. Rommel became a legend
and a hero to the men of the British Eighth Army. It did not affect
our morale nor our determination to fight, but he enjoyed an
affection with us that is probably unique in warfare . . . I never
saw him, but I learned in reading* The Rommel Papers *that I
was just a few hundred feet from him on more than one occasion.*

Peter ended the letter with his regrets that he could not attend the cer-
emonies over in Bath, so that Mrs. Rommel might sign his copy of Correlli
Barnett's *The Desert Generals*. His car, no longer the reliable Hillman Minx
but an army land rover, had a "mechanical defect."

Over the years, Pete often accompanied Jimmy on fishing trips.
Branch Lake remained one of their favorite spots. When Peter Godley
died in 2004, at the age of eighty-four, his cousin Jimmy was one of his
pallbearers.

◆

James Morrill remained working at Cap Morrill's until his retirement.
He never married. His spare hours were spent tinkering on cars when
he wasn't hunting or fishing. On the 40th anniversary of D-Day, the
local newspaper celebrated his bravery by doing a full-spread story.
Before that, Jimmy had said very little about his war experiences. Now,
in 1984, he remembered that in Plymouth, England, when they were being
fed so well before being shipped out to possible death, they were even given
ice cream.

The day they stormed Utah Beach, Jimmy and the other soldiers of the 4th
Infantry Division had learned that Brigadier General Theodore Roosevelt Jr.
would come ashore with them. The former president's son was walking with

a cane. He was already ill at fifty-six and suffering from a bad heart. The only general to make the landing with the troops, Roosevelt had to almost beg that he be given permission. He was right that it would be good for morale. "He was with us," Jimmy remembered proudly. Pushed 2,000 yards off their intended landing by strong ocean currents, Roosevelt scouted the terrain and uttered words that would become famous. "We'll start the war from right here!"

Morrill also recalled the day at Saint-Lô when General Lesley McNair was killed by fire from American bombers. Jimmy and his fellow soldiers had been crouching nearby. "I thought there was no end to the planes coming over," he said. "Everything we had was bombing Saint-Lô. I think only the cathedral was left standing. General Patton came in with his armor right after that. After that, we moved fast."

A month later, still in France, Teddy Roosevelt Jr. died from his bad heart. He was quickly buried at Sainte-Mère-Église. Honorary pallbearers had included generals Patton and Bradley. Among the few men chosen as the honor guard was Pvt. James Morrill, the small kid from Brewer, Maine. He assumed they picked men fast, given the circumstances, and he happened to be standing there. He felt privileged.

In 1984, Jimmy and brother Henry planned to attend an anniversary ceremony in Normandy. Presidents Reagan and Mitterrand would be there, as well as Queen Elizabeth II. But Jimmy had to cancel in order to undergo a planned heart surgery. When told the surgery would be scheduled for June 6, he smiled.

"Do you remember D-Day?" he asked the young doctor.

"No, I'm afraid not," was the reply.

"Well, I went through the whole nine yards and never got a scratch," Jimmy told him. "And now *you're* gonna mark me up?"

He died at home in 2010, being cared for by Cecile, his French-speaking sister-in-law who was Richard's widow. Almost ninety-one years old, Jimmy was last of the four Morrill brothers to pass away.

◆

Falling victim to time and neglect, Buonriposo, the Italian-styled villa overlooking the ocean in Bar Harbor, was demolished a decade after the 1952 northeaster. A new residence later replaced it. But rubble from the home where Edith and Ernesto Fabbri summered, along with his brother, Alessandro, often surfaced on the beach below. Years later, one resident reported finding fragments in the sand, such as a terra cotta angel and a belt buckle engraved with the name *Edith*.

◆

Hazel Tardiff was required to stay in the hospital for a full week, which was not uncommon then for new mothers. It was on Saturday that Phil came to tell his wife that her father had died suddenly. Having grown up close to him on the island, she was inconsolable. George Coombs had gone out that morning to shovel snow in his backyard and was soon feeling ill. He came in and sat in his rocking chair to drink a glass of water and baking soda that his wife, May, had mixed for him. When May returned a few minutes later to check on her husband, George had died with the glass still in his hand.

Isle au Haut had never left the former lobsterman's memories, especially those mornings when the sun rose yellow over Rich's Cove. That's when May, the orphan girl from San Francisco, was up early baking a week's pastries for the family. George had been born in 1884, just as the rusticators were discovering the island. He was sixteen years old that December day when his brother Herman drowned in the ocean with cousin Augustus. His mother, Julia Rich, had run across the island to the wharf, hoping to find her son still alive. It had been George, on a later December day, who carried his own child's coffin down the road to Coombs Cemetery.

Dying four days after the snow had stopped falling, George Grover Cleveland Coombs, age sixty-seven, was not counted among the fatalities of the 1952 northeaster. Still recovering from childbirth, Hazel could not convince Dr. Virginia Hamilton or the hospital staff to let her attend her father's funeral.

On Tuesday, a week after the birth of their last child, Hazel and her husband drove their new son, Dennis George, home to Varney Mill Road.

Before her death in 2006, Hazel Coombs Tardiff took her grown daughters to visit the place of her childhood memories, the old school, the church, the lighthouse, the seals at Boom Beach. She showed them the dance hall, once written up in an 1881 magazine as having the best floor in the state of Maine. And told them how she styled her hair for dances by holding a curling iron over the oil lamp. They went on the mailboat from Stonington to the place Champlain had named "High Island" back in 1604. Hazel took her girls to Isle au Haut.

◆

William "Bill" Dwyer was worried when he turned on his radio and heard that yet another possible northeaster might hit coastal Maine. He still privately hoped that Snooky would come home. He had grown tired of hearing his well-meaning neighbors tell him to just get another cat. There were barns sitting along country roads full of unwanted mousers, they assured him. One of them had to be yellow.

"Another cat at my age?" Bill had answered, time and again. "I'm nervous buying green bananas."

The truth was that he had grown to love the yellow cat that turned up one rainy day at his back door, looking skinny and ragged. He liked having it curl next to him on the sofa as he read the paper. Or on the end of his bed at night, warming his feet. He enjoyed sharing breakfast with another living thing when he walked into the kitchen each morning. But a week had passed since he last saw Snooky, that first night of the northeaster. If another storm was on its way, it was time to accept that his cat was gone.

On the morning of February 27, ten days after the snowstorm had begun, Bill's elderly neighbor two houses down on Bath Street asked for help in shoveling out her car. Because her family had delivered groceries during the snowfall, her vehicle had sat with snow piled up past its tires. Now she needed it for errands. Bill began shoveling slowly, mindful of what could happen to man who would celebrate his sixty-eighth birthday that summer. As he scooped heavy snow from beneath the front bumper, he thought he heard a soft meow. He shoveled more quickly, finally creating a tunnel beneath the undercarriage. And there crouched an emaciated Snooky.

"Snooky boy," Bill said as the cat emerged, blinking, into the light. He had had snow for water, but no food for almost ten days. Bill carried him home in his arms. In the kitchen, he put a small bit of tuna fish on a plate and poured some milk in a bowl. He held the cat back so it couldn't eat too much, too quickly.

"You beat Lazarus by six days, Snooky Boy," Bill told the cat as he stroked its back.

A journalist from the *Bath Independent* came out later that day when a neighbor of Bill's phoned with the exciting news. Snooky the cat had survived the storm.

Bill Dwyer, the former volunteer call man for Engine No. 4, died three years after the northeaster, just missing his seventy-second birthday in June. He always proudly said that he and Bath Iron Works had been born together in the summer of 1884.

There is no record, nor enough poetic license in the world, to guess when Snooky passed away.

◆

Civilization is a stream with banks . . . on the banks,
unnoticed, people build homes, make love, raise children,
sing songs, write poetry . . . The story of civilization
is the story of what happened on the banks.

—WILL DURANT, *The Story of Civilization.*

It's impossible to know how many northeasters have prowled the eastern coastline of this country during the vast span of unrecorded history. Or in the centuries since Samuel de Champlain sailed along its rugged edges. Before early newspaper reports or personal correspondences, storms were anonymous. They came and they went. The National Weather Service did not begin officially tracking them until its inception in 1890. So we will never know how many were successful in reaching landfall and inflicting serious damage on the landscape and wildlife, or later on inhabitants.

It's also impossible to know with certainty what the future holds. But according to statistics, the sea level has risen a foot since the big storm that locked Mark Twain in his New York hotel room. And it will rise four feet by the year 2100, more than tripling the rate of coastal flooding across the Northeast. Since the 1952 northeaster that took the lives of two men near the very spot where Maine lobsters were first discovered centuries earlier, the frequency and strength of these storms during the colder months have increased. Whether northeasters and blizzards are growing stronger or not, they are happening over an ocean that is much fuller than it was just a hundred years ago. This will undoubtedly mean dangerous flooding up ahead.

But human beings will respond to calamity as they have since the first shovel made from an oxen's shoulder blade began moving snow to make their lives better. With the James Webb telescope presently peering through the glittery ocean of our universe to study the first galaxies and stars, ordinary people, people who are sometimes unintentionally heroic, will confront tempests here on earth. They will trudge through the deep snow to carry injured children and pregnant women to hospitals. They will make biscuits and chowders for the living. They will lay fresh flowers and whisper prayers for the dead. They will continue adding chapters to that vast story of civilization, rebuilding on the banks of our shared stream those houses that have been swept away by floods, snows, and fires. They will sing songs and write poetry. They will persevere. In a brief matter of time, history will forget them.

What remains unchanged over the ages is that all storms are fickle. Having picked up power off Cape Hatteras on the Outer Banks of North Carolina, the northeaster that weathermen predicted would again hit New England made other plans. As Maine residents in small towns and coastal villages slept, the storm changed direction two hundred miles south of Cape Cod. Before dawn broke, it had moved out to sea, finally disappearing over gray water and ocean mist.

BIBLIOGRAPHY

BOOKS

Ashley, Alta. *Time Before the Boat: A Collection of Letters About Island Life.* Monhegan, Me.: Grey Gull Publications, 1985.

Banks, Ronald F., ed. *A History of Maine: A Collection of Readings on the History of Maine 1600–1976.* Dubuque, Ia: Kendal Hunt Publishing, 1976. (Banks reproduced this from Winship, George Parker. *Sailors Narratives of Voyages Along the New England Coast, 1524–1624.* Boston: Houghton-Mifflin, 1905.)

Bergreen, Laurence. *As Thousands Cheer: The Life of Irving Berlin.* New York: Viking, 1990.

Bianculli, Anthony J. *Trains and Technology: Cars.* Newark: University of Delaware Press, 2001.

Billings, Cathy. *The Maine Lobster Industry: A History of Culture, Conservation & Commerce.* Charleston, Sc: History Press, 2014.

Brechlin, Earl. *Wild! Weird! Wonderful! Maine.* Yarmouth, ME: Islandport Press, 2020.

"The Brownville Sesquicentennial Book 1824–1974." *The Piscataquis Observer* (1974).

Champlain, Samuel de. *The Voyages of Samuel de Champlain, 1604–1618.* New York: Charles Scribner's Sons, 1907. https://archive.org/details/voyagessam 00chamrich/page/n9/mode/2up.

Eskew, Garnett Laidlaw. *Cradle of Ships: A History of the Bath Ironworks.* New York: G. P. Putnam's Sons, 1958.

Fuller, John L. and Gerald P. Cooper. *A Biological Survey of the Lakes and Ponds of Mount Desert Island, and the Union and Lower Penobscot River Drainage Systems.* Augusta: Maine Department of Inland Fisheries and Game, 1946.

Gerrish, Judson and Henry. *The Brownville Centennial Book 1824–1924.* Dover-Foxcroft, Me: F. D. Barrows, 1924.

Lee III, James J. *U.S. Naval Radio Station, Apartment Building.* Historic Structure Report, Acadia National Park, Bar Harbor, Me. National Park Service, U.S. Department of the Interior, 2009. otter_cliffs_naval_radio_hsr.pdf.

Lewis, Susan M. *The Handbook of Brownville History.* Bangor, Me: Jordan-Frost
    Printing Co., 1935.

McLane, Charles B. *Islands of The Mid-Maine Coast: Penobscot and Blue Hill Bays.*
    Woolwich, Me: Kennebec River Press, 1982.

Ogilvie, Elisabeth. *My World Is an Island.* New York: McGraw-Hill, 1950.

Owen, Henry Wilson, A.B. *The Edward Clarence Plummer History of Bath, Maine.*
    Bath, Me: Times Company, 1936.

Phaneuf, David. *Winter Harbor's Naval History 1917–2002.* West Gouldsboro, Me:
    self-published, 2015.

Sawtell, Bill and Ruth Cyr. *History and Stories of Milo Junction/Derby, Maine.*
    Garland, Me: Moosehead Communications, 2010.

Snow, Edward Rowe. *Great Gales and Dire Disasters.* New York: Dodd, Mead and
    Co., 1952.

Snow, Edward Rowe. *Storms and Shipwrecks of New England,* Carlisle, Ma:
    Commonwealth Editions, 1943.

Thatcher, James. *History of the Town of Plymouth; From Its First Settlement in 1620,
    to the Year 1832.* Boston: Marsh, Capen & Lyon, 1832. https://archive.org
    /details/historyoftownofp02thac.

Thoreau, Henry David. *Walden.* New York: Dover, 1995.

ACADEMIC PAPERS

Ballantyne, Coco. "Hypothermia: How Long Can Someone Survive in Frigid
    Water?" *Scientific American.* (Jan. 16, 2009). https://www.scientificamerican
    .com/article/airplane-1549-hudson-hypothermia/.

Bryce-Smith, Roger and H. S. Davis. "Tidal Exchange in Respirators." *Semantic
    Scholar.* (1954). https://www.semanticscholar.org/paper/Tidal-exchange-in
    -respirators.-Bryce-Smith-Davis/b5879ff2a6ab2c8a8bd16086ae30ca1ecb0a2e
    24#citing-papers.

Safar, P. "From Back-Pressure Arm-Lift to Mouth-to-Mouth Control of Airway,
    and Beyond." In Safar, P. and J. O. Elam, (eds), *Advances in Cardiopulmonary
    Resuscitation.* New York: Springer, 1977. https://doi.org/10.1007/978-1-4612
    -6338-8_41.

Specht, Heinz, PhD. "Back-pressure Arm-lift Artificial Respiration." *Public Health
    Reports*, April 1, 1952, vol. 67, issue 4, 380–383. https://www.jstor.org/stable
    /10.2307/4588083?origin=crossref.

WEBSITES

Acadia National Park. "Kronprinzessin Cecilie." https://www.visitacadia.com
    /discover/kronprinzessin-cecilie/.

Ashley, Alta. Monhegan Memorial Library history. 1984. https://monheganlibrary
.com/history-of-the-library/.

Aviation Safety Network. "Sunday February 17, 1952." https://aviation-safety.net
/database/record.php?id=19520217-0.

Bahns, Ernst. "The Evolution of Ventilation." *Dragerwerk AG*. 10–11. https://ia
903208.us.archive.org/6/items/manual_Drager_Evolution_of_Ventilation
/Drager_Evolution_of_Ventilation.pdf.

Bahns, Ernst. "It Began with the Pulmotor—One Hundred Years of Artificial
Ventilation." www.draeger.com 2007. https://www.frca.co.uk/documents
/100%20YEARS%20VENTILATION%20BOOKLET.pdf.

Blanchard, Laura Jarrett. "Deadliest American Disasters and Large-Loss-of-Life
Events." http://www.usdeadlyevents.com/1778-dec-26-27-blizzard-new
-england-esp-ma-and-ri-brig-arnold-sloop-stark-151/.

Brandenburg, Katie. "Hearses Served as Ambulances at One Time." *Daily News*,
Bowling Green, Ky, Sept. 13, 2012.

Brooks, Ron. "How to Run in a Following Sea." LiveAbout.com.
Updated May 13, 2018. https://www.liveabout.com/how-to-run-in
-following-sea-2928429.

Browne, Patrick. "The Grim Fate of the Privateer 'General Arnold.'" *Historical
Digression* (blog), Oct. 25, 2011. https://historicaldigression.com/2011/10/25
/the-grim-fate-of-the-privateer-general-arnold/.

Cromarty, Ashley. "A History of Snow Plows & Snow Plowing." *Ezine Articles*,
July 23, 2009. https://ezinearticles.com/?A-History-of-Snow-Plows-and
-Snow-Plowing&id=2653088.

Dill, D. B. "Manual Artificial Respiration." *U.S. Armed Forces Medical Journal*.
(Feb. 3, 1952): (2):171–184. PMID: 14893592.

Emerson, Brad. "Italian Villas on the Maine Coast: Buonriposo." *Downeast
Dilettante* (blog). https://downeastdilettante.blogspot.com.

"The Evolution of CPR Training." *Emergency First Response*. https://www
.emergencyfirstresponse.com/the-evolution-of-cpr-training/.

Harvard Health Publishing Staff. "Protect Your Heart When Shoveling Snow."
Harvard Health Publishing, Harvard Medical School. Feb. 28, 2020. https
://www.health.harvard.edu/blog/protect-your-heart-when-shoveling
-snow-201101151153.

History.com. "*A Taste of Lobster History*." Aug. 22, 2018. https://www.history.com
/news/a-taste-of-lobster-history.

History.com. "Great Blizzard of '88 Hits East Coast." November 13, 2009. https
://www.history.com/this-day-in-history/great-blizzard-of-88-hits-east-coast.

"History of the Shovel." https://shovels.weebly.com/sources.html.

The History of the Shovel. *Rocketswag.* http://www.rocketswag.com/gardening
     /garden-tools/shovel/The-History-Of-The-Shovel.html.
Maine, An Encyclopedia. "Weather, Severe Events." https://maineanencyclopedia
     .com/weather-severe-events/.
Maine Organic Farmers and Gardeners. https://www.mofga.org/.
Miller, Tom. "The Ernesto and Edith Fabbri Mansion—No. 7 East 95th Street."
     Daytonian in Manhattan (blog). https://daytoninmanhattan.blogspot
     .com/2016/08/the-ernesto-and-edith-fabbri-mansion-no.html.
Mills, Mara. "Hearing Aids and the History of Electronics Miniaturization." *IEEE
     Annals of the History of Computing.* (2011): 33.2: 24–44. https://nyuscholars
     .nyu.edu/en/publications/hearing-aids-and-the-history-of-electronics
     -miniaturization.
National Climate Assessment. "Explore the Effects of Climate Change." https
     ://nca2014.globalchange.gov/.
Naval Security Group Activity, Winter Harbor. Wikipedia. https://en.wikipedia
     .org/wiki/Naval_Security_Group_Activity,_Winter_Harbor.
New England Historical Society. "Remembering the Great Snow of 1717 in New
     England." https://www.newenglandhistoricalsociety.com/great-snow-1717/.
New England Historical Society. "The Portland Gale of 1898, and the Cat that
     Saved a Life." https://www.newenglandhistoricalsociety.com/the-portland
     -gale-of-1898-and-the-cat-that-saved-a-life/.
NSGA Winter Harbor Maine. https://www.navycthistory.com/winterharbor
     _intro_main.html.
Rickard, H. J. "A New Method of Manual Artificial Respiration for Infants
     and Small Children." *Journal of the American Medical Association.* (Oct. 22,
     1955):159(8):754–765. doi: 10.1001/jama.1955.02960250016005. PMID:
     13263118.
Snow, Case. "Life Without Snow Plows: The History of Snow Removal."
     Encyclopedia of Snow Removal, Volume XI. Free Republic. Oct. 7, 2015.
     https://freerepublic.com/focus/f-chat/3921163/posts.
Sparks Journal 8-2. "Down East" Edition. *The Fabulous Radio NBD.* https://www
     .navy-radio.com/commsta/otter/NBD-Sparks-Vol8No2.pdf.
Winston, Jay S. "Monthly Weather Review February 1952; The Weather and
     Circulation of February 1952." Extended Forecast Section, U.S. Weather
     Bureau, Washington. https://citeseerx.ist.psu.edu/viewdoc/summary
     ?doi=10.1.1.394.7330.
"Worst One We Ever Saw Here: Remembering the Blizzard of 1952." *Mr.
     Lakefront* (blog). https://blog.mrlakefront.net/.

PERIODICALS

Barry, Dan. "At Howard Johnson's, a Final Few Scoops of Pistachio." *New York Times*, April 23, 2005. https://www.nytimes.com/2005/04/23/nyregion /at-howard-johnsons-a-final-few-scoops-of-pistachio.html.

Burnham, Emily. "164 Years Ago This Bangor Priest Was Tarred, Feathered, and Ridden on a Rail." *Bangor Daily News*, June 19, 2018. https://www .bangordailynews.com/2018/06/19/news/bangor/164-years-ago-this-bangor -priest-was-tarred-feathered-and-ridden-on-a-rail/.

Curtis, Abigail. "Whales, Furs, Fish and History: Monhegan Marks Capt. John Smith's 1614 Settlement." *Bangor Daily News*, July 28, 2014. https://www .bangordailynews.com/2014/07/28/news/whales-furs-fish-and-history -monhegan-marks-capt-john-smiths-1614-settlement/.

"German Gold Ship to Quit Bar Harbor: Kronprinzessin Cecilie Starts Today for Boston Under Escort to Face Libel Actions." *New York Times,* Nov. 6, 1914. https://www.nytimes.com/1914/11/06/archives/german-gold-ship-to-quit-bar -harbor-kronprinzessin-cecilie-starts.html.

"German Wireless in Maine Woods?" *Bangor Daily News*, October 1914.

Gold, Robert. "Pendleton Hero Bernie Weber Dies in Florida." *Cape Cod Times*. Jan. 25, 2009. https://www.capecodtimes.com/story/news/2009/01/26 /pendleton-hero-bernie-webber-dies/52120040007/.

"Hamilton Hopes Taft Snows Foes as Maine Snowed John Hamilton." *Portland Press Herald*, Feb. 19, 1952.

Howard, Alexander. "Hearing Aids: Smaller and Smarter." *New York Times*, Nov. 26, 1998. https://www.nytimes.com/1998/11/26/technology/hearing -aids-smaller-and-smarter.html.

Imhoff, Ernest F., and Frederick N. Rasmussen. "The Lost Art of Riveting." *Baltimore Sun*, Nov. 20, 1997. https://www.baltimoresun.com/news/bs-xpm -1997-11-20-1997324040-story.html.

Kimmelman, Michael. "Andrew Wyeth, Painter, Dies at 91." *New York Times*, Jan. 16, 2009.

Shaw, Dick. "D-Day, 40 Years After Landing in Normandy, the Horrible Reality of War Remembered." *Bangor Daily News*, June 2, 1984.

Knaak, Frederic W. "The Meuse-Argonne Offensive of 1918: Sacrifice and Triumph Against All Odds." *Star Tribune* (Minneapolis), Sept. 14, 2018. https://www.startribune.com/the-meuse-argonne-offensive-of-1918-sacrifice -and-triumph-against-all-odds/493347671/.

Latourneau, Gene. "Sportsmen Say." *Portland Press Herald,* March 1993. https ://digitalcommons.portlandlibrary.com/news_pph/1759.

Marsh, John F. "Help Wanted, Male—Long Hours—Hard Work." *Maine Fish and Game* (fall issue, 1971): vol. XIII, no. 4. https://issuu.com /mainestatelibrary/docs/13-4_-_maine_fish_and_game_magazine/30.

Owen, Joseph. *On This Date in Maine History: Feb. 10, 1886.* https://www .pressherald.com/2020/02/10/on-this-date-in-maine-history-feb-10.

Reilly, Wayne E. "German Wireless Sought in Maine Woods During First World War." *Bangor Daily News*, Jan. 18, 2015.

Sargent, Dave. "L-A Theaters of Old Were Grand Halls." *Lewiston Sun Journal* (Lewiston, Me), Jan. 24, 2012. https://www.sunjournal.com/2012/01/24 /l-a-theaters-old-grand-halls/.

Shea, Jim. "Blizzards: By Any Tally 1888 Is First." *Hartford Courant*, Jan. 23, 2014. https://www.courant.com/courant-250/moments-in-history/hc-250 -snowstorms-blizzard-weather-htmlstory.html.

"Tell 'Em About Taft." *Portland Sunday Telegram*, Feb. 17, 1952.

Weaver, Jacqueline. "Retired Cryptologist Traces Former Naval Base's History." *The Ellsworth American*, Aug. 1, 2015. https://www.ellsworthamerican.com /featured/retired-cryptologist-traces-former-naval-bases-history/.

Wilcox, Emily. "Naming the General Arnold's Lost Sailors." *Boston Globe*, Aug. 21, 2008.

Newspapers accessed from February 11, 1952, to February 27, 1952:

Advertiser-Democrat (Norway, Me); *Bar Harbor Times* (Me); *Bath Brunswick Times Record* (Me); *Bath Daily Times* (Me); *Bath Independent* (Me); *Boston Advertiser* (Ma); *Boston American* (Ma); *Boston Daily Record* (Ma); *Boston Traveler* (Ma); *Bridgton News* (Me); *Brunswick Record* (Me); *Caledonian Record* (St. Johnsbury, Vt); *Concord Enterprise* (Ma); *Courier-Gazette* (Rockland, Me); *Fitchburg Sentinel* (Ma); *Galveston Daily News* (Tx); *Hartford Courant* (Ct); *Lexington Herald* (Ma); *Lowell Sun* (Ma); *Newport Daily News* (Ri); *Pensacola News Journal* (Fl); *Piscataquis Observer* (Dover-Foxcroft, Me); *Portland Press Herald* (Me); *Portsmouth Herald* (Nh); *Providence Journal* (Ri); *Richmond Times-Dispatch* (Va); *Roanoke Times* (Va); *Springfield Union* (Ma); *Wilkes-Barre Times Leader* (Pa); and *Il Giornale d'Italia (Rome, Italy)*.

LETTERS AND ARCHIVES

Mark Twain Papers, Bancroft Library, Room 475, University of California, Berkeley.

Speed, Daniel. Personal diary entries, sent by daughter Rebecca Speed Randall.

YEARBOOKS

Portland High School, 1935.

Higgins Classical Institute Yearbook, Charleston, Me, 1946.

Thomaston High School Yearbook, 1941.

Thomaston High School Yearbook, 1943.

MAPS

Isle au Haut and Stonington, USGS, topographical quadrangle map, 1904.

Muscongus Bay, NOAA nautical chart, #13301, 2018.

Sanborn Fire Insurance Map from Bath, Sagadahoc County, Me. Sanborn Map
    Company, Dec. 1919.

Sanborn Fire Insurance Map from Warren, Knox County, Me. Sanborn Map
    Company, Dec. 1911.

PHOTO CREDITS

Thanks to the families of Harland and Alice Davis, Hazel Tardiff, Peter Godley, Jimmy Morrill, Paul Delaney, Westin Gamage, Dr. Charles North, Marden family and friends. Thanks also to Earl Brechlin for Buonriposo, and Paul Werner for the Herbert L. Douglas blizzard photos. Kerry E. Nelson collection for the Gannett photo of Paul Delaney. Marla H. Davis and the Mid Coast Hospital, Brunswick. The Brownville-Brownville Junction Historical Society's Parish House Museum. The Warren Historical Society. Cathryn Czajkowski and Nicole Luongo Cloutier, in Special Collections at the Portsmouth Public Library. The Maine Lumber camp postcard is courtesy of Special Collections, Raymond H. Fogler Library, University of Maine. John Hilling's painting Burning of the Old South Church, courtesy of the Jonathan and Karin Fielding Collection. The Howard Johnson postcard is courtesy of Collections of Boston Public Library. Thanks to the Library of Congress for access to photos that are public domain.

ENDNOTES

1   A blizzard has specifics, as defined by the National Weather Service. It is a sustained wind or frequent wind gusts at thirty-five miles an hour or greater, and coupled with considerable falling and/or blowing snow that frequently reduces visibility to less than a quarter mile. These conditions must prevail for a three-hour period or longer. The wind did not begin until after the snow was done falling for this February 11, 1952 storm. Therefore, it had blizzard-like conditions, but was not categorized as a blizzard. A northeaster is discussed in Part Two.

2   Born in British Guinea to a Guyanese mother of African descent and a father who was a wealthy Scottish plantation owner, Orr had been shipped off to Scotland at age three. Since anti-Catholicism had been erupting in occasional violence in the United Kingdom since the reign of Henry VIII, it's likely that this is where he picked up his religious intolerance.

3   The Tardiffs actually lived in what was and is considered North Bath.

4   With a nineteenth-century market demand for these products, shipments began to cities and towns in other states that were soon constructing their impressive museums, government buildings, libraries, and post offices that would last over time. Brownville's slate won first prize at the 1876 Centennial Exposition in Philadelphia, where brand-new concepts like Alexander Graham Bell's telephone, the Statue of Liberty's right arm and torch, and a bottle of Heinz tomato ketchup were showcased to a curious world.

5   The sisters considered booking sail on the *Titanic*. The daily newspapers had been filled with descriptions of the world's largest sailing ship, with its reputation for being unsinkable. But it would not sail from Southhampton until later that spring. The Burnett women left in January on the SS *Ascania* only to learn later, as the world did, that the *Titanic* went to its watery grave in April.

6   These storms are typecast according to the work done by an atmospheric
     scientist named James E. Miller, whose expertise involved storm formation
     in the Atlantic coastal region. Miller came up with a classification system in
     1946 that meteorologists refer to when categorizing northeasters. The storm
     described above is a classic Miller Type-A, with its origins off the coast of
     Georgia or South Carolina, although they can also begin down in the Gulf
     of Mexico. A Miller Type-B is a northeaster that gathers momentum in the
     Midwest and then moves eastward to the Appalachian Mountains. Once it
     hits those elevated peaks it loses its low-pressure center for the time being. But
     the low will redevelop once the mass reaches the East Coast. It then begins its
     journey northeast along the coastline as a classic Type-A would do.

7   Pontefract is one of the few places in England where licorice plants thrive
     in the deep and sandy soil. A market town since the Middle Ages, it has
     manufactured for three hundred years the small licorice confection famously
     known as a Pontefract cake.

8   The first report of the Surgeon General's Advisory Committee on Smoking
     and Health was released on January 11, 1964, by Surgeon General Luther
     Terry, MD. The biomedical literature had already published more than 7,000
     articles that concerned smoking and related diseases. The committee thus
     concluded that cigarette smoking is (1) a cause of lung cancer and laryngeal
     cancer in men; (2) a probable cause of lung cancer in women; and (3) the
     most important cause of chronic bronchitis. Ranked as one of the top news
     stories of 1964, the report's release made newspaper headlines and dominated
     television newscasts for several days.

9   One was defected American William Colepaugh who spoke no German, the
     other German-born Erich Gimpel. Operation Elster (Magpie, in English) was
     to gather and transmit back to Germany any technical data on the Allied war
     effort, once they built a radio they could use for these transmissions.

10   A famed raconteur, John Hamilton entertained folks with tales of Wendell
     Willkie, Henry Cabot Lodge, Thomas Dewey, and his famous client Harry
     Gold. But a more titillating story came with the second of his three wives, the
     famously beautiful debutante Jane Kendall Mason. She was a noted big game
     hunter with an ability to outdrink the best of men, many of whom were her
     lovers. She and her first husband, heir to a family fortune, adopted two boys
     when it was discovered Jane could not bear children. Finding herself void of
     maternal instincts as well, she left them to be raised by staff. It was reported
     that, perhaps to show her motherly affection, Jane Kendall shot a zebra foal
     and had it made into a rocking horse for the boys. She was reputed to have
     already killed the foal's mother, a white zebra so rare it was one of the few in

existence. One of Jane's many lovers was the African big game hunter Baron Bror von Blixen-Finecke, married to Karen Dinesen who had published her memoir *Out of Africa* in 1937. Jane and Ernest Hemingway were assumed lovers—Papa bragged that she had once crawled through his hotel window five flights up in order to avail herself of his masculinity—and he used her as his model for the callous Margot Macomber in his biting 1937 story "The Short Happy Life of Francis Macomber." John Hamilton fell madly in love with her. They divorced five years later.

11   Cora, the countess of Grantham in the hugely successful British drama *Downton Abbey*, mentions (Season 5, Episode 3) having a fitting with Edward Molyneux as the reason for her trip to London.

IN MEMORIAM
EDGAR COMEE (1917–2005)

I must mention the late Edgar A. Comee, of Brunswick, Maine, a man I wish I had known. He declared himself chairman of the "Ad Hoc Committee for Stamping Out Nor'easter," believing it to be an affectation of "non-sailors who wish to appear salty." If you read Comee's obituary in the *Portland Press Herald*, you will see that his eighty-eight years of life read like an adventure story.

The Oxford English Dictionary gives the term *nor'easter* an origin as early as the 1500s, long before any New Englander read a weather report. I have never used the term myself because I thought it sounded like how Hollywood has Mainers talk in movies. When I discovered the late Mr. Comee, I found vindication. At least I knew I wasn't alone. I have already been asked many times why I titled my book *Northeaster,* instead of *Nor'easter.* Unless readers see this writing, I will be asked it many more times.

The same year he died, Edgar Comee ended up in *Talk of the Town,* in the *New Yorker*: "The use of *nor'easter* to describe a northeast storm is a pretentious and altogether lamentable affectation, the odious, even loathsome, practice of landlubbers who would be seen as salty as the sea itself." I agree with Mr. Comee, and wish he were here so that we two could go for drinks at some oceanside café. We could sit and stare out to sea, saying nothing, but feeling our shared aversion to that term. Not only do Mainers *not* talk like the actors in *Murder, She Wrote*, northern Mainers don't even talk like southern Mainers. Therefore, it's *northeaster* as the title of this book.

Mr. Comee, may you rest in peace.

SPECIAL THANKS

I must say more about the children, relatives, and friends of the principals, some of whom answered dozens of questions and found old documents and photos as I worked on the book:

Bill Wilson, who was born a few months after his father, Harland Davis, died at sea. Bill and his wife **Sandy Bailey Wilson** came all the way north to Allagash so we could meet. He brought with him a box of memorabilia. And it was **Galen Wilson**, who remembered his mother Alice Davis Wilson in wonderful ways that included her cream puffs. And to **Michele Alice Ryan**, who was named for her grandmother. Michele sent me the postcards Harland mailed to Alice before he died in the northeaster. Michele also shared many photos left by her mother, Carolyn, the little girl Harland had accepted as his own.

Sue Godley (daughter of Peter Godley) and **Mary Tardiff Wirta** (daughter of Hazel Coombs Tardiff) who now feel like old friends. It was in 2004 when I first saw online photographs of the 1952 northeaster. Hazel was more than willing to talk to me. Being a novelist by trade, I was still not ready to take on a book of this genre. Another fifteen years would pass. How I wish I had met Hazel. She died in 2006. Through her daughter, I feel I came to know her. I also must thank her sons **Dennis Tardiff**, who was the soon-to-be-born baby who rode to the hospital on the toboggan, and **David Tardiff**, her oldest son, for their memories of growing up on Varney Mill Road in North Bath.

Barbara Ann Haigh. Two people kept me awake many nights during the nearly three years I worked on this book. One was Harland Davis, who appeared now and then in my dreams. The other was Barbi Haigh, who was still so young when she lost her parents. Barbara Ann, the adult woman who turned out amazingly well with her own family and career, shared with me the most heartfelt stories of loss. And yet she did it without remorse or regret. Instead, she was pleased to remember the two people she loved most back then, her parents Jimmy and Ellie Haigh. During the writing of this book, Barbi reconnected by telephone with Phyllis Sanders, widow of Earle Sanders, whose family took in Skybow, her beloved dog. (Barbara gave me permission to spell her nickname as "Barbie" for the book.)

Kathleen Delaney McNamara, daughter of Paul Delaney. Again, when I first thought I might write this book back in 2004, I found a phone number for Paul V. Delaney in Staten Island and dialed it. Paul had just passed away days earlier. But I spoke to his wonderful wife, Patricia. "We heard about his being trapped in the car many times over the years," she told me. Sadly, Patricia passed away in 2015, four years before I settled down to write the book. Thank you, Kathleen, for sharing your own memories of your father. (For reasons of clarity, I named Paul's sister Sarah. Her name is actually Ellen.)

Gerry Gamage and **David Gamage**, who gave me insight on the character of their father, boatswain's mate first class Weston Gamage, Jr. I remember the day when Adebo Adetona, Archives Specialist at the National Archives in Washington, D. C., found the 1952 report filed by Weston Gamage. It detailed the attempt by the coast guard to rescue two men from the sea. Gerry read the report and then wrote, "I can't believe Dad went back in that storm to find the boat." Sadly, David passed away before reading the finished book.

Ron Knowles, who now lives in Milo, and **Susan Worchester**, director of the Brownville-Brownville Junction Historical Society's Parish House Museum. Ron was a childhood friend to Ray "Sonny" Pomelow. (As a matter of fact, Sonny died on Ron's fifteenth birthday.) Because Ron retired years ago from a career as a trainman, I sent him dozens of questions about the trains running through Maine in the 1950s. I will treasure the map he drew for me of Brownville Junction to explain the layout of the tracks where the plow train hit Daniel Speed's car. Ron Knowles also introduced me to **John Ekholm**, whose memories of his friend Sonny Pomelow I shared in this book. Ron and I met in person in October 2022, after two years of emails and a few phone talks. We went together to visit Sonny Pomelow's grave. John Ekholm and Susan Worchester came with us. Susan also found photos and documents over the months and, like Ron, read my text pages for corrections. They both answered too many questions about Brownville and Brownville Junction in 1952. And thanks to Dorothy Perkins Gray for her memories of Stickney Hill and the train whistle that called kids home at night.

Bette Stubbs, niece of Ray "Sonny" Pomelow, for memories of her mother, Louise Pomelow Joslyn. It was Bette who learned about the light Louise saw in her bedroom the night before Sonny died. She kindly shared it with me, just in time to get it into this book. And thank you to Bette's daughter, **Julia Skidgel**.

Gordon Josyln, the late nephew of Ray "Sonny" Pomelow, who was sitting in the same room of the house where Sonny was waked when we talked on the phone. A veteran of the Vietnam War, Gordon passed away in 2021. And thanks to his wife, Nancy Joslyn.

Rebekah Speed Randall, for sharing pages with me from her father's diary. In those pages, I learned much about the personable man who was Daniel Speed. And thank you to his son, **Fred Speed**, for suggesting that I should "call Becky."

Anne B. Benaquist, for reading the many newspaper clippings we found about her mother, Dr. Virginia Clay Hamilton. And for adding her own memories of a remarkable woman who was ahead of her generation.

David Barstow and **Suzanne Crocket** for helping with the memory of their grandfather Dr. Charles North. And thanks to Suzanne's husband, **Charles Crocket**. Learning that Dr. North was partial to sky-blue DeSotos once he was forced to give up his horse and buggy is a gem any writer would cherish.

Darlene Page and **Jeanne Morrill**, Jimmy Morrill's and Peter Godley's nieces, for their encouragement. And thank you, **Cecile Pelletier Morrill**, that "French-speaking girl" from up north in Fort Kent, who married Richard Morrill, and later took care of brother-in-law Jimmy before he died.

Debbie Hogan Albert, for her help with tracking down info on game wardens. Her father was Robert Hogan, who went with Raymond Morse to search for the fishermen on Branch Lake. Debbie also found former railroad men to answer my questions.

ACKNOWLEDGMENTS

Tomislav Julian Viorikic, my husband of 30 years, at least in *this* crazy life.

Kerry E. Nelson, of West Bath, ME, my "indentured assistant," whose nineteen-year-old mother was in a Portland hospital undergoing emergency heart surgery during the 1952 northeaster. Kerry spent hours searching census reports and old records for names, dates, and unexpected gems to help me rebuild the lives of these real people, and resurrect the towns where they lived. She found people who knew and worked with Bill Dwyer, and that was invaluable to me. Her knowledge of local history wormed its way into the book wherever I could use it. And then, she and I went down a million rabbit holes together, reading about people long lost to time who had nothing to do with the book, but stoked our insatiable curiosities. Thank you, Yoda.

Paul Lucas, my agent at Janklow & Nesbit, for being a champion for the book.

Jessica Case, my editor at Pegasus Books, for her support and patience. The team at Pegasus: **Nicole Maher**, publicist; **Maria Fernandez**, interior design; **Drew Wheeler**, copyeditor; and **Mike Richards**, proofreader.

Lily St. Amant, Isabelle St. Amant, Dori Borkholder St. Amant, and **Kirk St. Amant**, in memory of my sister Joan.

Rosemary Kingsland, the late British author and my dear friend, who read my first rough chapters. I miss her friendship and her writerly input daily.

Note: *I am indebted to a small army of librarians, archivists, and historical societies who were quick to help. Most answered a river of questions pertaining to their expertise, from aquatic and marine biologists to funeral directors, from lobster fishermen to ice fishermen, from railroad men to emergency room doctors, from egg farmers to hair salon owners, from vintage car and radio enthusiasts to former firemen, from coast guard researchers to WWI and WWII historians. And many folks on Facebook who were extremely helpful. I hope I remember them all. There is no logical order or geography to these listings. But my gratitude is collective.*

Joe Cupo, retired Maine weatherman, who encouraged me to find 1952 weather surface maps for him to study. And who did his best to explain the characteristics of a northeaster to me. If anything is amiss in this book pertaining to weather, it won't be Joe Cupo's fault. It will be mine for not getting it down correctly. A natural teacher, Joe should explain the weather to all of us.

Matthew Scott, aquatic biologist (Emeritus AFS, AIFRB, and NALMS) who knows Branch Lake, and who answered many questions about lake waters and which fish are caught. He even answered this: "Is there a rise in the road, or was there a way, in 1952, where a driver could see headlights coming from the Narrows?" Matt sent detailed responses to each of my questions.

Karin E. Larson, historian at the Warren Historical Society, for numerous questions she answered and documents she found about the town of Warren, ME, in 1952. And thanks also to Warren Historical Society members **Sandra Overlock** and board member **Howard Wiley**. A big thanks also to **Robert "Bob" Wiley**, who shared, in a telephone interview, his memories of George Aspey and the Aspey family.

Mel Allen, who cares about writers and writing. He's the kind of friend we all wish we'd known since college.

Dr. James Harris, who was there to answer emails asking particulars about the death certificates and medical reports regarding several of my main characters. As if he isn't busy enough.

Kristen Carlson-Lewis, Isle au Haut Historical Society, for reading my manuscript for corrections. And to former president of the society and local historian **Harold Van Doren**, for the same reason. These two addressed many questions about their island, including, "Did the sun rise over Rich's Cove?" and "Where would George Coombs have carried his child's coffin in December, 1900?" Also, thank you to current president, **Tom Guglielmo**.

Ron Gamage, at Hall Funeral Home in Thomaston, ME, for searching old records to find details of Harland Davis's funeral, and Jimmy Haigh's being at the home. Ron got way too many questions from me and yet answered all with patience.

Robin Paquet, office manager at Purdy Funeral Service, Dover, NH. (This was formerly Wiggin Funeral Home which performed services for James "Jimmy" Haigh.) Robin found old documents, and asked questions of a former employee during the time James Haigh was brought down from the funeral home in Thomaston, ME.

Jim Sanders, son of Earle Sanders, was helpful in providing information. His father, Earle, was employed by Jimmy Haigh before his marriage to Phyllis Hoyt. Jim still runs what is now Sanders Fish Market, in Portsmouth, NH.

Phil Morrill, nephew of Jimmy Morrill, for information on the history of the Morrill family and Cap Morrill's tavern. The business has become Cap Morrill's Seafood, Inc. in Brewer, ME. Thanks **Dale** and **Gail Morrill** and **Joe Godley**, son of Peter Godley.

CWO Paul Roszkowski (U.S. Coast Guard Motion Picture, Television, and Author Liaison Office) who withstood a barrage of questions from me for over two years. Paul, who was a consultant for the film *The Finest Hours*, suggested many helpful archives and sources; William H. Thiesen, Ph.D., Atlantic area historian, United States Coast Guard; Joanie Gerin, archivist, National Archives at Boston; CDR William McKinstry, commanding officer for the CG-IMAT; The U.S. Coast Guard at Jonesboro and Boothbay Harbor, both in Maine; Sector Northern New England, South Portland, ME; Ben Critchley, retired coastguardsman; and Joseph Scherr, in New Hampshire.

Allison Wilson of Port Clyde, and his daughter Sandi Ochs, now of Racine, WI. Allison was most helpful in my figuring out what radios were in use in 1952. He spent forty-five years doing ship's pilot work and was a great source of information on ship-to-shore communication. And thanks to Dan Morris of Port Clyde, for putting me in touch.

Sandra Overlock, for connecting me with Trampus Copeland, and thus his mother, Carolyn Copeland. Carolyn is the daughter of Mary McLain, the girl Harland Davis loved in high school, according to the 1941 *Sea Breeze* yearbook. Carolyn Copeland still owns the hope chest Harland gave her mother, thinking they would marry one day. She had never been told what boyfriend had given it to Mary, or why the young couple broke up.

Ralph R. Garber for the phone interview. Ralph was stationed at the Winter Harbor Naval Base around the same time as Paul Delaney. Ralph took his future wife to the movies at the Criterion theater and then to Harris's Soda Shop after for burgers and cokes. And thanks to Michele Garber, his daughter, who helped deliver my questions. I must thank once again Adebo Adetona, archives specialist, National Archives, Washington, D. C.; Eric L. Annis, Lary Funeral Home, and John and Lyn Sherburne, for memories of Dr. Carde; Richard Sawtell; Jim Melanson; Galen Larsen; Eva Cushman, who just turned 98 this year and who told me her memories of Harland Davis. And her daughter Deborah Cushman Grace for helping with the questions. Also Tina Marriner for putting me in touch, and Sarah Parks of Rocky Face, GA; Larry Wells and Kathleen Woodruff Wickham, for reading an early draft; Earl Brechlin, of Bar Harbor, ME, for also reading; Carl Hileman of Tamms, IL, for the photo help; Ryan D. Pelletier and Chris Gardner, for putting me in touch with former coastguardsman Ben Critchley; Sandra Rayne Ph.D., physical scientist, NOAA's National Centers for Environmental Information (NCEI), for those surface weather maps; Amy Reytar, Textual Reference Archives II, National Archives at College Park, MD (for searching for files that recorded the railroad accident with Daniel Speed); Cathy Billings, Lobster Institute associate director, University of Maine; Richard A. Wahle, director, Lobster Institute and research professor, School of Marine Sciences, University of Maine; **Jon Bowdoin**, for a variety of info; **Renée DesRoberts**, special collections librarian, McArthur Public Library, Biddeford, ME; **Mia Boynton** at Monhegan Memorial Library for answering many questions; **Sherman Stanley, Jr**; **Richard Farrell**, at Tribler Cottage, on Monhegan Island; **Kerry Hardy, Jean English, and Jennifer**, of the Maine Organic Farmers and Gardeners Association, who fielded my question as to what pear tree was growing near Harland Davis's wharf, mentioned in Elisabeth Ogilvie's book. (I think it was decided it had escaped from cultivation.) **Sue Denison**, Norway Museum & Historical Society, Norway, ME; **Robin A. S. Haynes** and **Peter Goodwin**, former managers, and **Jill Piekut Roy**, archivist & special collections librarian, Sagadahoc History & Genealogy Room, Patten Free Library, Bath, ME; **Susan French Creamer Seigars**, of China, ME, for the old photos and Davis ancestry info; **Donna Peterson**, register of probate, Piscataquis County, Dover-Foxcroft, ME; **Wayne Barter**, Isle au Haut, ME; **Paul A. Werner**, West Bath, ME, who shared his postcards of the blizzard; **Reference Department staff**, Portland Public Library, Portland, ME; **Lawrence Renaud**, fire chief, Bath Fire Department, for checking the 1952 logs, and **Norman Kenney**, former fire chief, for his memories of William Dwyer, Bath, ME; **Donna E. Waterman**, for her memories of the Ice Follies trip to Boston, West Bath, ME.

    Samuel N. Howes, archivist III, Maine State Archives, Augusta, ME; **Kristie Spaulding**, main office secretary, Brunswick High School, Brunswick, ME; **Desiree Butterfield-Nagy**, UMaine Special Collections & Archives, Raymond H. Fogler Library, Orono, ME; **Sonja Plummer Morgan**, at the Mark and Emily Turner Memorial Library in Presque Isle, ME (and in memory of **Norma McEntee**); **Michele L. Brann**, Reference Services, Maine State Library, Augusta, ME; **Harley Brann**, Office of Chief Medical Examiner, Augusta, ME; **Sharon Thomas** and **Douglas Noble**, at the Baldwin Historical Society, Baldwin, ME; **Chad Poitras**, at Chad E. Poitras Cremation & Funeral Service, Buxton, ME; **Chuck Colbert**, production/house manager, The Criterion Theatre, Bar Harbor, ME; **Peter McKenney**, Great Falls Model Railroad Club, Auburn, ME; **Bob Bittner**, WJTO 730 AM, The Memories Station, West Bath, ME; **Nicole Pappaioanou**, Records Department, Portsmouth New Hampshire Police Department; **Eric Brooks**, curator/site manager, the Henry Clay Estate, Lexington, KY; **Melissa Martin**, reference librarian, Mark Twain Papers, The Bancroft Library, University of California/Berkeley; **Matthew Schultz**, Doylestown, PA, for his help with vintage snow plows used by the B & A in the 1940s and 1950s; **Brian Lehan**, corporate archivist, for the information on the Howard Johnson restaurant's cupola and weather-vane; **Beau**, in the Special Collections Department at Toronto Public Library, for info on the dance troupe that Gladys Adams Dwyer joined; **Tracy S. Skrabut**, archives technician, National Archives at

Boston, Waltham, MA; **Boston Public Library**; **Sarah Chapman**, librarian, Center for Homeland Defense and Security, Monterey, CA; **The Jesup Memorial Library** in Bar Harbor, particularly old friends **Kathleen C. Woodside, Melinda Rice**, and **Ruth Everland; Nicole Seavey**, police dispatcher, and **Karen Richter**, administrative assistant, Bar Harbor Police Department; **Edwin "Lee" Garrett**, for Bar Harbor info; **James Mondor**, Saco, ME; **Patricia Goodwin Berry, Julia Berry Proctor, Roberta Marden, Ted Taylor**, and **Janet Taylor McDaniel**, whose relatives were the Mardens who owned the Marden farm, and who shared their memories; **Ella Mae Grant Blackmore** and her sister **Peggy Grant**. Their mother was the nurse, Pearl Grant, who went to the Marden Farm to tend to Charles Voyer; **Anatole Brown**, education and program manager, and **Carolyn Parsons Roy**, collections and archives manager, at Dyer Library/Saco Museum; **Eric Donald Uldbjerg; Jennifer LeComte; Kevin McPherson; Camden Taylor; Patty King** and **Jessie Blanchard**, reference librarians at Rockland Public Library; **Kennebunk Historical Society; Kevin Norsworthy**, State Theatre, Port City Music Hall, Portland, ME; **Jim Pate**, Dennett, Craig & Pate Funeral Home, Saco Buxton, ME; **Mark Hutchins**, A.T. Hutchins Funeral and Cremation Services, Portland, ME; **David Jones** at Jones, Rich & Barnes Funeral Home; **Peter Scontras; Rosemary Martin Rovillard; Allison Braley**, granddaughter of Arthur LeBlanc, who was the helpful chef at the Howard Johnson's. And her mother **Ellen LeBlanc Almquist**, and her aunt **Susan LeBlanc Malley**, his daughters. Thank you, Susan, for checking my facts. **Bryan Kimball**, his son-in-law, and **Mark Dymkoski**, who worked at the Howard Johnson's and had high praise for both chef LeBlanc and then manager James P. Ivers; **Donna Thompson; Dale Woodward**, who has written about Winter Harbor Naval Station; **Pearl Barto**, Winter Harbor Historical Society; **Anne Marie V. Quin**, who is always ready to help; **Jeffrey Pellicani**, for telling me about ice expansion on lakes; **Dr. Charles E. Burden; Donald Spear; Dusty Innes; Paul Hutchins; Rick Bailey; Sharon Mercier; Jean Fullerton; Roslyn M. Hanna; Marion A. Beveridge, Dottie Burns**, and the late **Dorothy Marks**, who I almost got to interview; **Susie Nick**, Portland High School library and literacy specialist, who found photos and info on Hazel Coombs, in the PHS 1934 yearbook; **James "Jim" Goupee**, who was a good friend to Jimmy Morrill; **Laurie Goupee; Jody Gould; Virginia Vincent**, Warden Major's secretary, Maine Dept of Inland Fisheries & Wildlife; **Jonnie Maloney**, program coordinator, Lake Stewards of Maine; **Michele Daniels**, mayor of Brewer; **Samantha Giffard**, receptionist, Brewer Public Safety; **Deputy Chief Chris Martin**, Brewer Public Safety; **The Brewer Historical Society**, including the late **David Hanna**, and **Patricia Hanna**; and thanks, **Britney; Larry T. Doughty**, former Brewer city councilor; **Charlene Fox Clemons**, special collections/cataloger, Ellsworth Public Library; **Greta Schroeder**, former director at the Thompson Free Library in Dover-Foxcroft;

Janice Smith Johnsen and Betty Washburn Rosebush; Richard St. Amant for some trail lore; James and Kimberly Richards at Maine Wooden Buoys, Friendship, ME, for putting me in touch with Carl Simmons, who was then 95 years old and having two daily martinis with his friends. Carl remembered Harland dying in the northeaster, and shared his knowledge of lobster fishing in the early 1950s. (I have a martini raincheck.) Elizabeth Lewis, office manager, Dover-Foxcroft Police Department; the Piscataquis Police Department; the Brownville Town Office; Steve Heiny, director of camping, Bangor YMCA Wilderness Center at Camp Jordan, and also Justin Cullens; Donald Roger Dower, then of Canaan, ME, who was the three-month-old baby stranded with his parents at the Howard Johnson's. He was surprised I found him, and joked, "I don't remember anything." Eugene Dinsmore, who I spoke to by phone, and who was the seven-month-old baby scalded by coffee and whose father walked seven miles in the snow to get help; author Ken Hatchette, for information on the Canadian Pacific Railroad; Lynn Weston and Dan Peters, volunteers at Brownville-Brownville Junction Historical Society's Parish House Museum; Bill Sawtell, for his numerous railroad books, and his niece Shelly Sawtell, and also Xiaorong (Sharon) Horton; Leverett Fernald, for his interviews with former train men concerning trains circa 1952, and snowplow operations during their time of employment, with Robert Spencer, former brakeman and conductor, Rodney Stanhope, former engineer. and Ed Berry, former shop supervisor; and also, Charlie Freeman; Johanna Wolter (Dresden, Germany) who helped me track down the canna lily at Landesberg; Eric Anderson, president of the Cushing Historical Society, and Marie Spague, vice-president; Allen "Pat" and Nancy Malone; Peter Crooker; Scott Stanton; Sophie Poirier, at Galerie Cosner, in Montreal, Canada; Sue Podnar Barry and Susan Morrison, for helping with research; Maureen Hanley; Megan Hull; Benjamin Barr; Audrey Cannan; Terry Kelly; Carole and Bob Keyser of NY; Abe Miller-Rushing, at the Schoodic Institute; and Emanuel, Carmen, and Eva Zeries, of Romania, all for computer help.

And thank you to these friends for a variety of favors and encouragement: Kathy Kelly Rioux; Paula Dean Womack Mondelli; Kathleen Wallace King, Randy Ford, and Larry Wells, all three of who helped me keep Rosemary Kingsland's memory alive; Nicole Olivier; Allen Jackson; John Valeri; Deborah Joy Corey; Janet Mills; Tony Buxton; Charlie Miller; Jack Page; Gina Nadeau; Leslie Oster; Colleen and Rick McLaughlin; Darlene Kelly Dumond; Crystal Condo; Nelson and Mariana Eddy; Dr. Paul Gahlinger; Danna Brown Nickerson, Elizabeth R. Nelson, and Albert R. "Alb" Nelson, in memory of Melissa Smart Ferrucci; and to my mother-in-law Kristina Copkov, who is one of a kind.